OUTNUMBERED

Also available in the Bloomsbury Sigma series:

OUTNUMBERED

FROM FACEBOOK AND GOOGLE TO FAKE NEWS AND FILTER-BUBBLES – THE ALGORITHMS THAT CONTROL OUR LIVES

David Sumpter

BLOOMSBURY SIGMA

LONDON · OXFORD · NEW YORK · NEW DELHI · SYDNEY

BLOOMSBURY SIGMA
Bloomsbury Publishing Plc
50 Bedford Square, London, WC1B 3DP, UK

BLOOMSBURY, BLOOMSBURY SIGMA and the Bloomsbury Sigma logo
are trademarks of Bloomsbury Publishing Plc

First published in the UK in 2018

A catalogue record for this book is available from the British Library

Library of Congress Cataloguing-in-Publication data has been applied for

ISBN: HB: 978-1-4729-4741-3; TPB: 978-1-4729-4743-7;
eBook: 978-1-4729-4742-0

4 6 8 10 9 7 5 3

Illustrations by David Sumpter

Bloomsbury Sigma, Book Thirty-six

Typeset by Deanta Global Publishing Services, Chennai, India
Printed and bound in Great Britain by CPI Group (UK) Ltd, Croydon CR0 4YY

To find out more about our authors and books visit www.bloomsbury.com
and sign up for our newsletters

Contents

PART ONE
ANALYSING US

CHAPTER ONE

Finding Banksy

In March 2016, three researchers from London and a criminologist from Texas published a paper in the *Journal of Spatial Science*. The methods presented were dry and scholarly, but the article was not an abstract academic exercise. The title made the aim clear: 'Tagging Banksy: using geographic profiling to investigate a modern art mystery'. Mathematics was being used to track down the world's most famous graffiti artist.

The researchers used Banksy's website to identify the location of his street art. They then systematically visited each of his works, both in London and his home town of Bristol, with a GPS recorder. Data collected, they created a heat map in which warmer areas showed where Banksy was more likely to have lived, assuming that he created his work near to his home.

The hottest point on the London geo-profile map was only 500m from the former address of the girlfriend of one person who had previously been suggested to be Banksy. On the Bristol map, the colour was hottest around the house this same person lived at and the football pitch of the team he played for. The geo-profiled person, the article concluded, was very likely to be Banksy.

Upon reading the article, my first reaction was the mixture of interest and jealousy that most academics feel when their colleagues do something smart. This was a clever choice of application. Exactly the type of applied maths I aspire to do: well executed with a twist that captures the imagination. I wished I'd done it myself.

But as I read on, I started to feel slightly annoyed. I like Banksy. I have that book of his pictures and flippant quotes on my coffee table. I have wandered around the backstreets of cities looking for his murals. I have laughed at the video where his valuable artwork fails to sell from a stall he set up in Central Park in New York City. His creations in the West Bank and the migrant camps in Calais have provided me with an uncomfortable reminder of my own privileged position. I don't want some emotionally disconnected academics using their algorithm to tell me who Banksy is. The whole point of Banksy is that he sneaks unnoticed through the night and in the morning light his art reveals the hypocrisies of our society.

Mathematics is destroying art: cold logical statistics chasing hoody-wearing freedom fighters around the backstreets of London. It is just not right. It is the police and tabloid newspapers that should be looking for Banksy, not liberal-minded academics. Who do these smart-arses think they are?

I read the Banksy article a few weeks before my own book, *Soccermatics*, was due to be published. My aim in writing a football book had been to take readers on a mathematical journey into the beautiful game. I wanted to show that the structure and patterns underlying football were full of mathematics.

When the book came out, there was a lot of media interest in the idea, and every day I was asked to do interviews. For the most part, the journalists were as fascinated as I was by mathematical football, but there was a niggling question that kept being asked. A question they told me their readers would want to know the answer to: 'Do you think that numbers are taking the passion out of the game?'

'Of course they aren't!', I answered indignantly. I explained that there was always room for both logical thought and passion in the broad church that is football.

But hadn't the mathematical unmasking taken some of the mystery out of Banksy's art? Wasn't I cynically using football in the same way? Maybe I was doing exactly the same thing to football fans as the spatial statisticians were now doing to street graffiti.

Later that month, I was invited to Google's London headquarters to give a talk about the mathematics of football.

The talk was arranged by the book's publicist, Rebecca, and we were both eager to have a look around Google's research facilities.

We weren't disappointed. Their offices were at a well-appointed address on Buckingham Palace Road, where there were large Lego structures in the lobby and fridges stuffed full of health drinks and superfoods. The 'Googlers', which was how they referred to themselves, were clearly very proud of their surroundings.

I asked some of the Googlers what the company was up to now. I had heard about the self-driving cars, Google Glass and contact lenses, the drones delivering packages to our door, and the idea of injecting nanoparticles into our bodies to detect disease, and I wanted to know more about the rumours.

But the Googlers were cagey. After a run of bad publicity about the company becoming an over-creative hub for crazy ideas, the policy was to stop telling the outside world too much about what it was up to. Google's head of advanced technology projects at the time, Regina Dugan, had previously held the same position at the American government's Defense Advanced Research Projects Agency (DARPA). She had employed 'need-to-know' rules for information sharing to Google. The research division now consisted of small units, each of which worked on its own project and shared ideas and data internally within the groups.[1]

After some more quizzing, one of the Googlers finally mentioned a project. 'I heard we are using DeepMind to look at medical diagnostics around kidney failure,' he said.

The plan was to use machine learning to find patterns in kidney disease that doctors had missed. The DeepMind in question is the branch of Google that has programmed a computer to become the best go player in the world and trained an algorithm to master playing *Space Invaders* and other old arcade games. Now it would search through the UK's National Health Service (NHS) patient records to try and find patterns in the occurrence of diseases. DeepMind would become an intelligent computing assistant to doctors.

Just as when I first read the Banksy article, I again felt that pang of jealous but excited desire to be in the Googlers' position: to also have a chance to find diseases and improve healthcare using algorithms. Imagine having the finance and the data to carry out a project like this, and save lives, using maths.

Rebecca was less impressed. 'I'm not sure I'd like Google having all my medical data,' she said. 'Its kind of worrying when you think how they might use it together with other personal data.'

Her reaction made me think again. Comprehensive databases of both health and lifestyle data are accumulating faster than ever. While Google was following strict rules about data protection, the potential was still there. In the future, society might demand that we link up our search, social media and health data, in order to get a complete picture of who we are and why we get ill.

We didn't have much time to discuss the pros and cons of data-driven medical research before my presentation. And once I started talking about football I quickly forgot the whole issue. The Googlers were fascinated and there were lots of questions afterwards. What was the latest in state-of-the-art camera tracking technology? Could football managers be replaced by machine learning techniques that gradually improve tactics? And there were technical questions about data collection and robot football.

None of the Googlers asked me if I thought all this data would rip the soul out of the game. They would presumably have been very happy to connect up 24-hour health and nutrition monitoring devices to all the players to get a complete picture of their fitness levels. The more data the better.

I can relate to the Googlers, just as I can relate to the Banksy statisticians. It is very cool to have an NHS patient database on your computer, or to be able to track down 'criminals' using spatial statistics. In London and Berlin, New York and California, Stockholm, Shanghai and Tokyo, maths geeks like us are collecting and processing data. We are designing algorithms to recognise faces, to understand language and to learn our music tastes. We are building personal assistants and chatbots that can help you fix your computer. We are predicting the outcome of elections and sporting events. We are finding the perfect match for single people or helping them swipe through all available alternatives. We are trying to provide the most relevant news for you on Facebook and Twitter. We are making sure that you find out about the best holiday breaks and cheap flights. Our aim is to use data and algorithms as a force for good.

But is it really as simple as that? Are mathematicians making the world a better place? My reaction to Banksy's unmasking, the football journalists' reaction to my *Soccermatics* models and Rebecca's reaction to Google's use of medical databases are not unusual or unfounded. They are very natural. Algorithms are used everywhere to help us better understand the world. But do we really want to understand the world better if this means dissecting the things we love and losing our personal integrity? Are the algorithms we develop doing things society wants to do or just serving the interest of a few geeks and the multinational companies they work for? And is there a risk, as we develop better and better artificial intelligence (AI), that the algorithms start to take over? That maths starts to make our decisions for us?

The way in which the real world and maths interact is never straightforward. All of us, including myself, sometimes fall into the trap of seeing maths as turning a handle and getting results. Applied mathematicians are trained to see the world in terms of a modelling cycle. The cycle starts when consumers in the real world give us a problem they want to solve, be it finding Banksy or designing an online search engine. We take out our mathematical toolkit, power up our computers, program our code and see if we can get better answers. We implement an algorithm and supply it to the customers who asked for it. They give us feedback, and the cycle continues.

This handle-turning and model-cycling makes mathematicians detached. The offices of Google and Facebook, equipped with toys and indoor playgrounds, give their super-smart employees the illusion that they are in total control of the problems they are dealing with. The splendid isolation of university departments means we don't have to confront our theories with reality. This is wrong. The real world has real problems and it is our job to come up with real solutions There is so much more complexity to every problem than just making calculations.

During the months that followed my visit to Google in May 2016, I started to see a new type of maths story in the newspapers. An uncertainty was spreading across Europe and the US. Google's search engine was making racist autocomplete suggestions; Twitterbots were spreading fake news; Stephen Hawking was worried about artificial intelligence; far-right groups were living in algorithmically created filter-bubbles; Facebook was measuring our personalities, and these were being exploited to target voters. One after another, the stories of the dangers of algorithms accumulated. Even the mathematicians' ability to make predictions was called into question as statistical models got both Brexit and Trump wrong.

Stories about the maths of football, love, weddings, graffiti and other fun things were suddenly replaced by the maths of sexism, hate, dystopia and embarrassing errors in opinion poll calculations.

When I reread the scientific article on Banksy, a bit more carefully this time, I found that very little new evidence was presented about his identity. While the researchers mapped out the precise position of 140 artworks, they only investigated the addresses of one single suspect. The suspect in question had already been identified eight years earlier in the Daily Mail as the real Banksy.[2] The newspaper established that he was from a suburban middle-class background and wasn't the working-class hero we expect our graffiti artists to be.

One of the scientific article's co-authors, Steve Le Comber, was candid with the BBC about why they focused on the Daily Mail's suspect. He said: 'It rapidly became apparent that there is only one serious suspect, and everyone knows who it is. If you Google Banksy and [name of suspect] you get something like 43,500 hits.'[3]

Long before the mathematicians came along, the Internet thought it knew the true identity of Banksy. What the researchers did was associate numbers with that knowledge, but it isn't really clear what these numbers mean. The scientists had only tested one suspect in one case. The article illustrated the methods, but was far from conclusive proof that these methods actually worked.

The media wasn't too concerned with the study's limitations. A scandal story from the *Daily Mail* had now become a serious news item, covered by the *Guardian*, the *Economist* and the BBC. Maths provided legitimacy to a rumour. And the study contributed to the belief that the task of finding 'criminals' could be solved by an algorithm.

We change scene to a courtroom. Imagine that instead of standing accused of entertaining us with his revered street art, Banksy is a Muslim man accused of covering the walls of Birmingham with pro-Islamic State propaganda. Let's further imagine that the police have done a bit of background research and found that the graffiti started when their suspect moved to Birmingham from Islamabad. But they can't use this in court because they don't have any evidence. So what do the police do? Well, they call in the mathematicians. Applying their algorithm, the police's statistical experts predict with 65.2% certainty that a particular house belongs to Islamic Banksy and the anti-terror squad are called. A week later, under the Prevention of Terrorism Act, Islamic Banksy is placed under house arrest.

This scenario is not so far from how Steve and his colleagues envisage using their research results. In their article, they write that finding Banksy 'supports previous suggestions that analysis of minor terrorism-related acts (e.g. graffiti) could be used to help locate terrorist bases before more serious incidents occur'. With mathematical support in hand, Islamic Banksy is charged and convicted. What was previously a weak form of circumstantial evidence is now statistical proof.

That is just the start. After the success of identifying Islamic Banksy, private companies start competing for contracts to deliver statistical advice to the police force. After it secures its first contract, Google feeds the entire police records into DeepMind to identify potential terrorists. A few years later the government introduces the 'common sense' measure, supported by the general public, of integrating our web-search data with Google's police records database. The result is the creation of 'artificial intelligence officers' that can use our search and movement data to reason about our motives and future behaviour. Officer AI is assigned a designated strike force to carry out night-time raids on potential thought-terrorists. The dark mathematical future arrives at an alarming rate.

We are only a few pages in and maths is not just being a spoilsport, it is undermining our personal integrity, it is providing legitimacy to tabloid rumours, it is accusing the citizens of Birmingham of acts of terror, it is collating vast amounts of data inside massive unaccountable corporations and it is building a super-brain to monitor our behaviour.

How serious are these issues and how realistic are these scenarios? I decided to find out in the only way I knew how: by looking at the data, computing the statistics and doing the maths.

CHAPTER TWO

Make Some Noise

After the mathematical unmasking of Banksy had sunk in, I realised that I had somehow missed the sheer scale of the change that algorithms were making to our society. But let me be clear. I certainly hadn't missed the development of the mathematics. Machine learning, statistical models and artificial intelligence are all things I actively research and talk about with my colleagues every day. I read the latest articles and keep up to date with the biggest developments. But I was concentrating on the scientific side of things: looking at how the algorithms work in the abstract. I had failed to think seriously about the consequences of their usage. I hadn't thought about how the tools I was helping to develop were changing society.

I wasn't the only mathematician to experience this revelation, and compared with my somewhat frivolous concerns about Banksy's identity being revealed, some of my colleagues had found things to be really worried about. Near the end of 2016, mathematician Cathy O'Neil published her book, *Weapons of Math Destruction*, documenting misuses of algorithms in everything from evaluating teachers and online advertising of university courses to providing private credit and predicting criminal reoffending.[1] Her conclusions were frightening: algorithms were making arbitrary decisions about us, often based on dubious assumptions and inaccurate data.

A year earlier, Frank Pasquale, law professor at the University of Maryland, had published his book *The Black Box Society*. He argued that while our private lives were becoming increasingly open – as we share details of our lifestyle, our aspirations, our movements and our social lives online – the tools used by Wall Street and Silicon Valley companies to analyse our data were closed to scrutiny. These

black boxes were influencing the information we saw and
making decisions about us, while we were left unable to work
out how these algorithms operate.

Online, I found that a loosely organised group of data
scientists were responding to these challenges and analysing
how algorithms were applied in society.

These activists' most immediate concerns revolved around
the transparency and the potential for bias. While you are
online, Google collects information on the sites you visit and
uses this data to decide which adverts to show you. Search for
Spain and over the next few days you are shown holidays you
might want to take there. Search for football and you'll start
to see more betting sites appear on the screen. Search for links
about the dangers of black box algorithms and you'll be
offered a subscription to the *New York Times*.

Over time, Google builds up a picture of your interests and
classifies them. It is straightforward to find out what it has
inferred about you using 'ads settings' on your Google
account.[2] When I went into these settings, I found that
Google knows a fair bit about me: soccer, politics, online
communities and outdoors are all correctly identified as
things that I enjoy. But some of the other topics suggested
were a bit spurious: American football and cycling are two
sports Google thinks I like, but I have no real interest in. I felt
I had to set it straight. Inside ads settings, I clicked on the
crosses next to the sports I don't want to know about and then
added mathematics to the list.

At Carnegie Mellon University in Pennsylvania, US, PhD
student Amit Datta and his colleagues conducted a series of
experiments to measure exactly how Google classifies us.
They designed an automated tool that creates Google 'agents'
that open up webpages with predefined settings. These agents
then visited sites related to particular subjects, and the
researchers looked at both the adverts the agents were shown
and the changes in their ads settings. When the agents
browsed sites related to substance abuse, they were shown
adverts for rehab centres. Similarly, agents browsing sites
associated with disability were more likely to be shown

adverts for wheelchairs. Google isn't entirely honest with us, though. At no point were the agents' ads settings updated to tell the user the conclusions Google's algorithm had drawn about them. Even when we use our settings to tell Google which adverts we do and don't want to be shown, it makes its own decisions about what to show us.

Some readers might be interested to know that Google didn't change its advertising for agents that browsed adult websites. When I asked Amit if this meant users could freely search as much porn as they want without increasing the probability of an inappropriate ad popping up on their screens at another time, he advised caution: 'Google might change its advertising on other websites that we didn't browse. So, inappropriate Google ads may pop up on other sites.'

All the big Internet services – including Google, Yahoo, Facebook, Microsoft and Apple – build up a personalised picture of our interests and use these to decide what adverts to show us. These services are transparent to some degree, allowing users to review their settings. It lies in these companies' interests to ask us if they have properly understood our tastes. But they certainly don't tell us everything they know about us.

Angela Grammatas, who works as a programmer in marketing analytics, emphasised to me that there is little doubt that retargeting – which is the technical term for the online advertising that uses recent searches to choose what products to show users – is highly effective. She told me about how the Campbell's 'SoupTube' campaign used Google's Vogon system to show users specific versions of an advert that best matched their interests. According to Google, the campaign led to a 55 per cent increase in sales.[3]

Angela told me: 'Google is relatively benign, but the power of the "like" button on Facebook to target adverts is scary. Your "liking" gives a lot of insight into you as a person.' What worried Angela most was a change in the law in the US allowing Internet service providers (ISPs) – the telecommunication companies that deliver the Internet to your house – to store and use their customers' search history.

Unlike Google and Facebook, ISPs offer little or no transparency about the information they collect about you. They can potentially link your browsing history to your home address and share your data with third-party advertisers.

Angela was sufficiently concerned about this change in the law that she created a web browser plugin that prevents ISPs, or anyone else, from collecting useful data about their customers. She called the plugin Noiszy, because its role is, quite literally, to generate browsing noise. While she browses the sites she is interested in, Noiszy works in the background, browsing randomly from the top 40 news sites. The ISPs have no way of knowing which sites interest her and which don't. The change to the ads shown in her browser was immediately noticeable. 'Suddenly I was seeing all these adverts for Fox News … a big change from the "liberal media" bubble I had lived in previously,' she told me. Angela, who is happily married, also noticed a very large number of adverts for wedding dresses. Her browser didn't know who she was any more.

I found Angela's approach fascinating because she appeared torn on the issue of how companies use our data. Her day job was to create effective retargeting adverts. She was clearly very good at her professional work and believed that she was helping people find the products they wanted. But in her spare time, she had created a plugin that defeated these exact same advertising campaigns and had made her software freely available to anyone who wanted to use it.[4] 'If we all use Noiszy,' she wrote on the webpage accompanying the plugin, 'companies and organizations lose the ability to figure us out.' She told me that her aim was to increase understanding and debate about how online advertising worked.

Despite the apparent contradictions, I somehow understood the logic of Angela's approach. Yes, there were cases of blatant discrimination that needed to be detected and stopped. Certainly, some of the targeted advertising for short-term loans and dodgy 'university' qualifications was immoral.[5] And, yes, our web browser sometimes draws slightly strange conclusions about us. But usually the effect of re-advertising is relatively benign, and most of us don't mind being shown a few products we might be interested in. Angela is right to focus on educating us about how modern advertising works. It is our responsibility to understand the algorithms that try to sell us stuff and to make sure ISPs respect our rights.

The conclusions algorithms draw about us can be discriminatory, though. To investigate gender bias, Amit and his colleagues initialised 500 'male' agents (who had their gender set to male) and 500 'female' agents who browsed from a predefined set of job-related websites.[6] After these browsing sessions, they looked at the adverts shown to the agents. Despite similar browsing histories, the men were more likely to be shown a specific ad from the website careerchange.com with the headline: '$200k+ jobs – execs only'. Women were more likely to be offered adverts for generic recruitment sites. This type of discrimination is blatant and potentially illegal.

The president of the company that runs careerchange.com, Waffles Pi Natusch, told the *Pittsburgh Post-Gazette* that he wasn't sure how the ads ended up skewing so heavily towards men, but accepted that some of the company's ad preferences – for individuals with executive-level experience, over 45 years old and earning more than $100,000 per year – might push Google's algorithms in that direction.[7] This explanation was strange since the experimental agents differed only in gender, not in salary or age. Either Google's ads algorithm has either directly or indirectly made the link between men and high executive pay, or careerchange.com had inadvertently clicked a box that targeted their adverts at men.[8]

It was at this point that Amit and his colleagues' investigations ended. He told me that while they received no response from Google when they published their work, the Internet giant changed its interface so that they couldn't run their agent experiments any more. The black box was shut for good.

Julia Angwin and colleagues at non-profit newsroom ProPublica have opened up a large number of black boxes over the past two years, in a series of articles on machine bias. Using data collected from more than 7,000 criminal defendants in Florida, Julia showed that one of the algorithms widely used by the US judicial system was biased against African Americans.[9] Even accounting for criminal history, age, gender and future offences, they showed that African Americans were 45 per cent more likely to be predicted by the algorithm to be in a high-risk category for crime.

This type of discrimination isn't limited to the legal system. In another ProPublica study, Julia placed an advert on Facebook targeted at 'first-time buyers' and people 'who were likely to move', but which excluded those who had an 'ethnic affinity' for 'African American', 'Asian American' or 'Hispanic'. Facebook accepted and published the advert despite it being in violation of the US's Fair Housing Act.[10] Excluding certain groups, even if it is based on their 'affinity' (which Facebook measures by looking at the pages and posts users engage with) rather than their actual race, is discrimination.

The ProPublica journalists comprise part of a much larger movement of data journalists and scientists investigating these questions. Massachusetts Institute of Technology (MIT) graduate student Joy Buolamwini found that modern face recognition couldn't distinguish her face, so she started collecting a more ethnically diverse dataset of faces that could be used to train and improve future recognition systems;[11] Jonathan Albright at North Carolina's Elon University investigated the data used by Google's search engine to try to understand why its autocomplete often gave racist and offensive completions[12] and Jenna Burrell at the University of Berkeley, California reverse-engineered her email's spam filter to see if it explicitly discriminated against Nigerians (in this case it didn't).[13]

These researchers – together with Angela Grammatas, Amit Datta, Cathy O'Neil and many more – are determined to monitor the algorithms created by Internet giants and the security industry. They share their data and their code openly in online repositories so that others can download them and find out how they work. Many of them conduct their studies in their spare time, using their skills as programmers, academics and statisticians to find out how the world is being reshaped by algorithms.

Dissecting algorithms might not carry the street cred of Banksy's art, but in comparison to the closed view of the future and the secretive research units I had found in Google's London headquarters, I was deeply impressed by how these activists worked and provided their results for everyone to use.

This movement was having an effect. Facebook made changes so that the adverts like those placed by Julia Angwin were no longer possible. After an article in the *Guardian*, Google improved its autocomplete so that it no longer made anti-Semitic, sexist or racist suggestions. And although Google had been less responsive to Amit Datta's work, he has talked to Microsoft about helping them detect discrimination in online job advertising. Activism was starting to make a difference.

CHAPTER THREE

The Principal Components of Friendship

I'm perhaps not the typical activist type. I'm a professor of applied maths and part of the scientific establishment. I am a British middle-class, middle-aged father of two, who has escaped the political turmoil of his home country for a quiet life in Sweden. I contribute to algorithm development, which was why I was invited to talk at Google. Every day at work, I use mathematics to better understand our social behaviour, to explain how we interact with each other and to find the consequences of those interactions. But I don't make a lot of noise about political questions.

I am not proud of my inaction. Talking to Angela Grammatas and others like her made me feel that I was stuck with my head in my laptop, ignoring the problems. The rise of algorithms comes at a time of increasing uncertainty in Europe and the US. These changes make many people feel outnumbered. Nearly every news story – from Donald Trump's use of the political consultants Cambridge Analytica to target voters during his election campaign, to a failure of statisticians to predict the UK's Brexit vote – has an algorithmic angle. When I listen to my friends talk about these issues or follow discussions on Twitter, I find that I can't properly answer the questions that are raised. People want to know what is going on inside the black boxes that are used to assess and influence us.

The term 'black box' was used both by Frank Pasquale in the title of his book *The Black Box Society* and by ProPublica in their series of articles and video shorts on algorithms 'Breaking the Black Box'. It is a powerful image. You input your data, wait for the model to process it, and receive an answer. You can't see what is going on inside. Predicting criminal reoffending is carried out by a black box. Facebook and Google advertising is generated via a black box. Tagging Banksy is performed by a black box.

The black box imagery can give us a sense of helplessness, a feeling that we can't understand what algorithms do with our data. But this feeling can be misleading. We can and should look at what goes on inside algorithms. It was here I felt I could do something. I could look into the algorithmic black boxes used in our society and see how they work. I might not be much of an activist, but I could answer some of the questions people were asking about the changes in society.

It was time to get to work.

I thought about what Angela Grammatas had told me about Facebook: that it was the site that knew most about us. The social media giant was the best place to start investigating how algorithms classify us. I needed to start by looking at something I was sure I fully understood: my own social life. By creating a black box model of my own friends, I should be able to understand the steps taken by data scientists working at Facebook and Google. I would get first-hand experience of the techniques they use. The scale of my model would be much smaller, but the approach would be the same.

Angela was right. My friends' Facebook pages contain a massive amount of information about their lives. I open my Facebook news feed and see an update from a grumpy professor complaining because the driver of his train braked too quickly. I see photos scanned in and uploaded from school discos that took place 25 years ago. I see holiday snaps and after-work beers. I see jokes about Donald Trump, campaigns to improve healthcare and housing and outrage over political decisions. I see people boasting about their success at work and in parenting. I see wedding photos, pictures of small babies and children splashing happily in swimming pools. Everything from the deeply personal to the overtly political can be found on our Facebook news feed.

I take 32 of my Facebook friends and go through their 15 most recent posts.[1] I classify each of the posts as falling into one of 13 common categories: family/partner, outdoors, work, jokes/memes, product/advertising, politics/news, music/sport/film, animals, friends, local events, thoughts/ reflections, activism and lifestyle. I then make a matrix, a 32-row by 13-column spreadsheet, where I fill in the number of times my friends have made a particular type of post. For example, the row for one of my friends from university, Mark, has one post about his job, eight posts with pictures of his family holidays, three posts about the politics of Brexit (as a Scot living in Paris, he is against), one travel post from New York City and one post marking himself safe from the November 2015 terror attack in Paris. On the row for my colleague Torbjörn, the most common posts (five of them) are about the Nobel Prize dinner that he not only attended but was interviewed about for Swedish TV. I count these as work posts, along with two other posts about talks he has given. Torbjörn has two family posts and the remaining posts are spread among various other categories.

To start to get to grips with the balance between work and family life for Mark, Torbjörn and my other friends, I plotted the number of posts they made about work versus the number of family/partner posts. The result is shown in Figure 3.1. Mark is in the top left-hand corner, with eight family posts and one work post. Torbjörn is bottom, mid-right, with

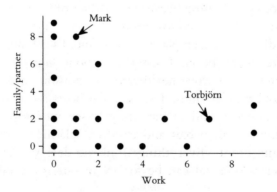

Figure 3.1 Classifying my friends on two dimensions of work and family/partner. Each dot represents how many times that particular friend posted on Facebook about each subject.

seven work posts and two family posts. Each of the other dots represents how my friends are positioned in these two dimensions of work and family/partner.

A small number of my friends can be categorised as posting mainly about work, while others can be viewed as family posters. But some of them post about both topics and there are also quite a few who don't post much about either. Each of the categories of posts can be thought of as a dimension of space, of which I've shown two: the first dimension is work posts and the second dimension is family/partner posts. But I could continue with a third dimension of outdoors posts, a fourth dimension of politics/news and so on. Each of my friends is a single point in this 13-dimensional space.

The problem is that as I increase the dimensions, the data becomes more difficult to visualise. I can't form a clear idea in my head of what a point in a 13-dimensional space looks like. Two dimensions, as in Figure 3.1, aren't a problem. Three dimensions are possible to deal with: first I imagine the points placed in a cube and then think about how the points would change position as I rotate it. But we just can't think about how the world looks in four or more dimensions. Our brains are limited to working in two or three because this is what we experience in everyday life.

For those of us who can't see points in four or more dimensions, the simplest way to navigate is by taking lots of two-dimensional snapshots. Figure 3.1 is a single snapshot of the relationship between work and family/partner. From other similar snapshots, I could see that it was unusual for those people who make lifestyle posts, about food and travel, to also post about politics/news. These two interests are negatively correlated: friends who liked to share pictures of a restaurant they'd just visited tended not to give their opinions on current affairs. Other types of post are positively correlated: my friends who wrote about music, films and sport also tended to share jokes and memes.

Comparing data in pairs starts to give us a feel for some of the patterns in a 13-dimensional data set, but it is not a particularly systematic approach. There are 78 pairs of relationships to look at,[2] so it takes time to plot them all and examine them. In some cases, there are multiple relationships: those people who shared jokes and memes, and wrote about music and films, also shared news and politics but tended not to write lifestyle posts about themselves. I would like to have a way of systematically ranking the strength of these relationships: to find out which are most important and that best capture the differences between my friends.

I applied a method known as principal component analysis, or PCA for short, to my friends' data. PCA is a statistical method that rotates my original 13-dimensional data set, where each post category is a single dimension, to reveal the most important relationships between posts. The first principal component, the relationship that gives the strongest correlations in the data, is a straight line that travels up through the family/partner, lifestyle and friends dimensions, while travelling down through the jokes/memes, politics/news and work dimensions. This is the most important relationship that distinguishes my friends. Some like to post about what they have been up to in their private lives, others like to share what has happened in the world and at work.

The second most important relationship in the data distinguishes work from hobbies and interests: passing up through work and lifestyle and down through music/sport/film, politics/news and other posts about culture. Mathematically, the second principal component is the line that lies closest to the data points while at a right angle to the first principal component. Drawing lines and rotating data in 13 dimensions is difficult for us to visualise, but it is straightforward to calculate the lines and perform the rotations required using a computer.[3] Figure 3.2 shows how the 13 different post types can be viewed in a two-dimensional space.

The largest positive contribution to the first principal component (shown on the right in Figure 3.2) comes from family/partner posts, the second largest is from lifestyle and the third and fourth are for friends and outdoors. The

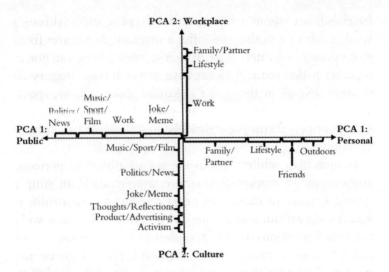

Figure 3.2 The first and second principal components from an analysis of my friends' posts. Left to right is the first principal component, which I label as 'public vs personal'. Up and down is the second principal component, which I label as 'culture vs workplace'. The size of the contribution (negative or positive) is indicated by the length of the segment of the line the component makes up. So, for example, 'family/partner' is the most prominent post for first principal component.

common theme of these posts is that they are all related to our personal lives – they involve writing about the things we do and the people we do them with. The categories jokes/memes, work, music, sport, film and politics/news give negative contributions to the same principal component (on the left in Figure 3.2). These types of posts are all related to public life: they are about our jobs or they comment on news or current events. I call this first principal component 'public vs personal' because it captures differences in how my friends use Facebook to either post about themselves or comment on the world at large.

The biggest positive contributing category to the second principal component is work, followed by lifestyle (see the upper line in Figure 3.2). Many of the lifestyle posts I see on

Facebook are about trips that friends have made through work – relaxing with a beer after a meeting or pictures from a conference dinner. So grouping these two categories together makes sense. The negative scores in this category all relate to events in the wider cultural sphere – news, sports and jokes all feature, as do activism and advertising. So this second principal component can be best described as 'culture vs workplace'.

Notice that, while I assign the names public vs personal and culture vs workplace to the components, I am simply giving a name to categories generated by the algorithm. It was the algorithm, and not me, who decided that these were the best dimensions on which to describe my friends.

Now these dimensions are defined I can categorise my friends. Which of them are more interested in public life or their personal lives? Which are more work or more culture-oriented?

To find out, I now place my friends in the public vs personal and culture vs workplace two-dimensional space (Figure 3.3). When I saw these names pop up on my screen,

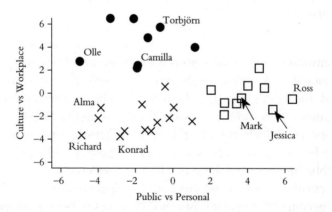

Figure 3.3 A breakdown of my friends along two principal components. The people on the right (squares) post mainly about friends, family and personal life. People at the bottom left (crosses) concentrate their posts on the news, sports and the public sphere. The people on the top left (circles) post mainly about work.

I knew immediately that the components made sense. Most of the people marked by squares on the far right – like Jessica, Mark and Ross – have children and are happy to share information about them. Most of the people in the bottom left corner, marked by crosses, didn't have children at the time I did my analysis and write more about current events: Alma about literature and the theatre, Konrad about computer games and Richard about politics. The top-left group, marked by circles, are typically academics writing about their work and recently published articles. Torbjörn is a fellow mathematical biologist. We'll meet Olle, a slightly eccentric mathematician from Gothenburg, who combines interests in politics with work posts, later in the book.

What surprised me most was the degree to which this classification captured genuine similarities and differences between my friends. Remember, I didn't tell the PCA algorithm how I wanted to categorise people. I provided a wide set of 13 categories that the PCA reduced to the two most pertinent dimensions: public vs personal and culture vs workplace. And these dimensions make sense – the most important differences between my friends do lie along these dimensions.

The grouping of my friends into three separate types (circles, squares and crosses) was also performed by an algorithm. I used a computational technique called 'k-means clustering' to group together individuals based on their distance from each other along the dimensions created by the PCA. This created three categories: people who use Facebook to focus on their private life (squares); people who focus on their work and work-related lifestyle (circles) and people who use Facebook to comment on wider events in society (crosses). Again, I had asked the algorithm to find the most efficient way to group my friends, and these were the categories it came up with. Principal component analysis uses the data to classify people, rather than relying on our preconceptions.

The friends I had classified largely agreed with the conclusions of my principal component analysis. Camilla, who I had down as a work person, said the analysis reflected the way she uses Facebook, mainly for information about working life. She uses other social media sites for friends and family. Ross uses it in the opposite way, 'just for a few family pics as your graph says', he told me.

Torbjörn didn't like that I had him down as 'just work and no fun', but agreed that on Facebook he was focused on the world of work rather than his private life.

Classifying my Facebook friends was fun, but there was a more serious side to reducing friendships to two dimensions. Principal component analysis and similar mathematical approaches underlie most of the algorithms used to classify our behaviour. This approach is used in models of criminal reoffending to turn questionnaire answers given by a defendant into predictions about whether he or she will commit further crimes. It is used by Twitter to work out how much you earn, it is used by Google to create advertising preferences. The amounts of data involved and the dimensions on which we are classified are much larger, but the approach is the same as I used: rotate and reduce until the algorithm starts to understand you.

It's amazing how just 15 posts can capture our lives. Imagine what Facebook can do with billions of posts …

CHAPTER FOUR

One Hundred Dimensions of You

Facebook has two billion users worldwide who make tens of millions of posts per hour, providing a vast record of our social activity. Professor Michal Kosinski at the Stanford Graduate School of Business was one of the first researchers to realise that we could use the PCA approach to categorise people on the basis of the vast quantities of data they were uploading to social media. While a PhD student at the University of Cambridge, he set up the myPersonality project together with David Stillwell. They collected a remarkable data set: over three million people gave them permission to access and store their Facebook profiles. Many of these people then took a battery of psychometric tests, measuring intelligence, personality and happiness, and answered questions about sexual orientation, drug use and other aspects of their lifestyle. The data provided Michal with a massive database that connects how the things we write, share and like on Facebook relate to our behaviour, our views and our personality.

Michal started by looking at properties where we can be categorised in one of two ways: Republican or Democrat, gay or straight, Christian or Muslim, male or female, single or in a relationship and so on. His research question was whether our 'likes' can be used to assess who we are: which 'likes' are most likely to be associated with each category?

The table providing some of the predictive 'likes' in Michal and his colleagues' scientific article is a list of painfully embarrassing stereotypes.[1] In 2010/11, when the study was carried out, gay males liked Sue Sylvester from the TV show *Glee*, Adam Lambert from *American Idol* and supported various human rights campaigns. Straight men liked Foot Locker, Wu-Tang Clan, the X Games and Bruce Lee. People with only a few friends liked the computer game *Minecraft*, hard rock music and 'walking with your friend and randomly pushing them into someone'. People with lots of friends liked Jennifer Lopez. People with low IQ liked the National Lampoon's character, Clark Griswold, 'being a mom' and Harley Davidson motorbikes. People with high IQ liked Mozart, science, *The Lord of the Rings* and *The Godfather*. African Americans liked Hello Kitty, Barack Obama and rapper Nicki Minaj, but were less keen on camping or Mitt Romney than other ethnic groups.

These observations don't mean that we should conclude a person is gay based on a single 'like' for Sue Sylvester, or that just because someone likes Mozart they are smart. That would be school playground logic: 'Ha ha, you like *Minecraft* … you don't have any friends.' Such reasoning is not only unpleasant, it is usually wrong.

Instead, Michal found that each 'like' provided a little bit of information about a person and an accumulation of lots of 'likes' allows his algorithm to draw reliable conclusions. To combine all of our 'likes', Michal and his colleagues used principal component analysis. He took all the tens of thousands of different 'likes' people had made and used PCA to find which of them contributed to the same component. For example, The Beatles, Red Hot Chili Peppers and the TV show *House* were all found together on one dimension, which we might label 'middle-aged rock music and films'. Another dimension, which we might label 'advertised products' included Disney Pixar, Oreo and YouTube. And so on. Michal found that between 40 and 100 dimensions were needed to accurately classify us.[2]

Michal emphasises that computers find more subtle relationships than humans. 'It is easy to infer that someone who goes to gay clubs and buys gay magazines is more likely to be gay. But computers can make those predictions on signals that aren't so telling to us,' he told me. Indeed, only five per cent of users labelled as gay by his algorithm liked an explicitly gay Facebook page. It was the combination of lots of different likes, ranging from Britney Spears to *Desperate*

Housewives that allowed the algorithm to determine the user's sexual orientation.

While Michal's large-scale analysis of data from Facebook is new, his use of principal component analysis is not. Over the past 50 years, sociologists and psychologists have used PCA to categorise our personalities, our societal values, our political views and our socio-economic status. We like to think of ourselves as multi-dimensional. We see ourselves as complicated individuals with many different sides to our characters. We tell ourselves that we are unique, that we are shaped by the millions of unique events that occur during our lifetimes. But PCA can reduce those millions of dimensions of complexity to a much smaller number of dimensions which can be used to put us in boxes or, to use a more appropriate visual metaphor, to represent us as a few different symbols. PCA tells me that I can, more or less, see my friends as clusters of circles, squares and crosses.

It is through this mission to view us as a cluster of symbols that we arrive at the list of personality traits that psychologists call the Big Five.

Psychologists' research into personality builds on our everyday understanding of our friends and acquaintances. We all know people who are friendly and talkative, people who prefer to be around others and like going out. We call these people 'extroverts'. We also know people who like reading and computer programming, who enjoy being alone and speak less often when they are in groups. We call these people 'introverts'. These are not just frivolous notions. They are correct and useful ways to describe people.

Many of our intuitions about other people lack scientific rigour, though. There are lots of words we can use to describe our friends and colleagues: argumentative, agreeable, competent, compliant, idealistic, assertive, self-disciplined, depressive, impulsive … the list goes on and on. Confronted with these long lists of adjectives and after conducting wide-ranging questionnaires in which respondents ranked themselves on the basis of a large number of different statements – such as 'I get chores done right away' and 'I talk to a lot of different people at parties' – psychologists turned to PCA to find the underlying patterns in our personalities. And the results have been startlingly consistent. Rotating through all the dimensions of human adjectives and independent of the types of questions posed, in most cases, psychologists recover the same Big Five components of personality: openness, conscientiousness, extroversion, agreeableness and neuroticism.[3]

These five traits are not arbitrary choices, but robust, repeatable measurements that summarise what it means to be human.

What occurred to Michal was that if the Big Five personality traits were robust and if Facebook likes could be used to predict IQ and political views, it should also be possible to predict personality from our Facebook profiles.

And it was. Outgoing people on Facebook like dancing, theatre and *Beer Pong*. Shy people like *anime*, role-playing games and Terry Pratchett books. Neurotic people like Kurt Cobain, emo music and say 'sometimes I hate myself'. Calm people like skydiving, football and business administration. Many stereotypes were confirmed by people's 'likes', but quite a few unexpected relationships also appeared. I am a relatively calm person, who is partial to a game of football, but I'm hardly going to throw myself out of a plane, with or without a parachute on. It is the accumulation of many different likes that reveals our personalities, not any single click.

We are clicking our personalities into Facebook, hour after hour, day after day. Smileys, thumbs ups, 'likes', frowns and hearts. We are telling Facebook who we are and what we think. We are revealing ourselves to a social media site in a level of detail that we usually reserve for only our closest friends. And unlike our friends – who tend to forget the details and are forgiving in the conclusions they draw about us – Facebook is systematically collecting, processing and analysing our emotional state. It is rotating our personalities in hundreds of dimensions, so it can find the most cold, rational direction to view us from.

Facebook's researchers have mastered the techniques of reducing our dimensionality. In the study I made of my own friends, I reduced 13 dimensions of posting by 32 people to two dimensions using an algorithm that took less than a second to execute. Michal used a similar algorithm and took about an hour to reduce 55,000 different likes by tens of thousands of people to the 40 or so dimensions required to predict their personalities. Facebook is working on a very different scale. Its current methods take less than a second to reduce one million different like categories, chosen by 100,000 different people to a few hundred dimensions.[4]

The methods employed by Facebook build on the mathematics of randomness. Rotating the data for one million different like categories one million times – which is the method that worked relatively quickly for my 15 categories – takes a long time. So Facebook's algorithm starts instead by proposing a random set of dimensions with which to describe us. The algorithm then evaluates how well these random dimensions perform, allowing it to find a new set of dimensions that improves its description. After only a few iterations, Facebook can have a pretty good idea of the most important components that describe its users.

While Facebook can reduce millions of likes to a few hundred components, it is difficult for us to visualise these components.[5] Our brains, which work in two or three dimensions and not in several hundred, reach their limits very quickly. So in order to help us understand how it sees us, Facebook has created names for the categories its algorithms come up with. To find out how the company has characterised you, you have to first log in, click on the drop-down option menu in the top right-hand corner and click on 'settings'. Once in settings, choose 'ads', click on the edit button to the right of 'ads based on my preferences' and then 'visit ad preferences'; finally, click the list 'lifestyle and culture'.

When the *New York Times* published an article telling people how to find these ad preferences, the paper's readers found all sorts of interesting categories assigned to them. These included people who Facebook categorised in terms of their interest in 'toast', 'tugboats', 'neck' and 'platypus'.[6]

I can see the humour in this and the people who found these categories had a good laugh about how Facebook had misunderstood them. Maybe it had. But when we see these categories, it is important to remember that they are an attempt to put words to a much deeper algorithmic understanding that Facebook has created of its users. The algorithms that categorise us don't rely on words. The words are put there to help us humans understand statistical relationships between peoples' interests. In fact, these relationships can't be expressed in words like 'toast' and 'platypus'. They can't really be explained in words at all. We simply can't get to grips with the high-dimensional understanding Facebook has created of us.

This was a point that Michal came back to repeatedly when I spoke to him. He emphasised that humans think about other people in just a small number of dimensions – age, race, gender and, if we know them a bit better, personality – while algorithms are already processing billions of data points and making classifications in hundreds of dimensions. When we don't understand how Facebook classifies us, the joke is on us, not on the algorithms. We no longer have the capability to fully understand the output of the algorithms we have created.

'We are better than computers at doing insignificant things that we, for some reason, think are very important, like walking around,' Michal told me. 'But computers can do other intelligent tasks that we can never do.' In Michal's view, principal component analysis is the first step towards creating a computerised, high-dimensional understanding of human personality that outperforms the understanding we currently have of ourselves.

Facebook has obtained a series of patents that allows the company to utilise its multi-dimensional understanding of us. One of the first of these patents was for romantic matchmaking.[7]

Facebook's idea is to find matches in users' profiles between friends of friends. We often imagine how our single friends, who might not know each other, might make a good couple. Facebook's system could make these suggestions for us, based on the personality it establishes in the users' profiles. The patent proposes allowing single users to look for friends of friends to 'identify [potential dates] that match a preference set of desired traits, interests, or experiences'. The mutual friend can then be asked if they are willing to be an intermediary.

If Facebook can find you a partner, then it should surely be able to find you a job. In 2012, researcher Donald Kluemper and his colleagues found that a human assessment of the Facebook profiles of 586 students (primarily white females) provided reliable assessments of their workplace hireability.[8] Several third-party companies have filed patents to use Facebook and other social-networking sites to extend this finding to automated job-matching.[9] The advantage for employers in using Facebook, over purely professional services like LinkedIn, is that your Facebook profile is (for better or worse) more likely to reveal the true you.

Facebook is also looking at ways of measuring your state of mind from your posts, your emotions from your facial expressions in photographs and your level of engagement from your rate of interaction with your screen.[10] Academic research has confirmed that these techniques can give some insight into our state of mind. For example, the speed with which users move their mouse during routine computer tasks can reveal the emotional content of what they are looking at on screen.[11] Principal component analysis can break down the way you are interacting with your phone or your computer, in order to build up a picture of how you are feeling.[12]

These developments suggest a future in which Facebook tracks our every emotion and continually manipulates us in our consumer choices, our relationships and our job opportunities.

If you regularly use Facebook, Instagram, Snapchat, Twitter or any other social media site, then you are outnumbered. You are allowing your personality to be placed as a point in hundreds of dimensions, your emotions to be enumerated and your future behaviour to be modelled and predicted. This is all done effectively and automatically, in a way that most of us can hardly comprehend.

CHAPTER FIVE

Cambridge Hyperbolytica

After the 2016 US presidential election, a company called Cambridge Analytica announced that its data-driven campaign had been instrumental in Donald Trump's victory. The front page of the company's website featured a montage of clips from CNN, CBSN, Bloomberg and Sky News showing the story of how it had used targeted online marketing and micro-level polling data to influence voters. The film ended with a quote from the political pollster Frank Luntz: 'There are no longer any experts except Cambridge Analytica. They were Trump's team who figured out how to win.'

Cambridge Analytica (CA) gave a great deal of prominence to the Big Five personality model in its promotional material. The company claimed to have collected hundreds of millions of data points about large numbers of US voters. CA claimed it could use this data to provide a picture of the voters' personalities that went beyond the traditional demographics of gender, age and income. Michal Kosinski, who led the study of Facebook personalities in Chapter 4, was very clear when I spoke to him that he had no involvement with CA, but he admitted that the company could use methods similar to those used in his scientific research in order to target voters. If CA had access to voters' Facebook profiles it could determine which types of advertising would have the greatest effect on them.

It is a scary thought. Facebook's data can be used to reveal our preferences, our IQ and our personalities. These dimensions could then, in theory at least, be used to target us with messages that appeal to us as individuals: a person with low IQ could be fed unverifiable conspiracy theories about Hillary Clinton's email accounts; a person with high IQ could be told that Donald Trump is a pragmatic businessman;

a person with 'African American affinity' (as Facebook calls it) could be told about inner-city rejuvenation; an unemployed white worker could be told about building a wall to keep out immigrants and 'Hispanic affinity' voters could be informed about a tough line on Castro's Cuba. Neurotic personalities can be targeted with fear, compassionate people targeted with empathy, and extroverts provided with a fun way to share the message.

Instead of focusing on a central message in the traditional media, the candidate in such a campaign might focus on discrediting journalists and news agencies, which are trying to form a global view of the election. With the mass media questioned, tailored messages would be delivered direct to individuals, providing them with propaganda that conformed to their already established world view.

At the time I started to look into Cambridge Analytica, in the autumn of 2017, the company was much more cautious of how it portrayed its role in Trump's victory. The *Guardian* and *Observer* newspapers had investigated several aspects of how CA collected and shared data, both in connection with the US presidential election and the UK vote to leave the European Union.[1] As a result, CA was now at pains to downplay its use of psychology in campaigning. The company profiled itself as using artificial intelligence to do audience segmentation, and no longer used the term 'personality'.

I contacted CA's public relations office several times to ask if I could speak to a member of technical staff about how the algorithms worked. The replies were polite, but for some reason, the person I should talk to was always 'away for the bank holiday' or 'on holiday now'. After a long row of excuses, my email requests were no longer answered.

So I decided to find out for myself how an approach to winning elections based on political personalities could work.

Before we get carried away with the idea of right-wing politicians tapping into a 100-dimensional representation of America's voters, we need to think about how accurately the dimensions inside a computer really represent us as people.

If I wanted to childishly insult a computer's ability to think, I might refer to the fact that it operates in binary: in terms of ones and zeros. But this is the wrong way to describe mathematical models. In fact, it is often humans that see things in binary states of black and white. We make statements like 'he is too stupid to understand', 'she is a typical Republican' or 'that person shares everything on Twitter' almost reflexively, without thinking about the lack of subtlety of our judgements. It is humans who see the world in binary.[2]

Well-designed algorithms seldom categorise events as belonging to one of two categories. They work instead with rankings or probabilities. The Facebook personality model assigns an extrovert/introvert ranking to each user or gives the probability of a user being 'single' or 'in a relationship'. These models take a range of factors and produce a single number that is proportional to the probability of a particular fact being true about the person.

The most basic method used for converting large numbers of dimensions to a probability, or ranking, is known as regression. Statisticians have used regression models for over a century, with applications starting in biology and expanding to economics, the insurance industry, political science and sociology. A regression model takes the data we already have about a person and uses them to predict something we don't know about him or her. To do this – a process known as fitting the model – we first need to have a group of people about whom we already know the thing we are trying to predict.

Consider the relationship between age and voting for Brexit. Ten days before the vote on whether or not the UK should leave the EU, YouGov conducted a poll in which it asked people how they would vote. There were four different age groups included in the survey: 18–24, 25–49, 50–64 and 65+, and the answer the respondents gave depended on their age group. Figure 5.1 shows a regression model I fitted to the voters' intentions. As age increases so does the probability that the person will vote to leave the EU.

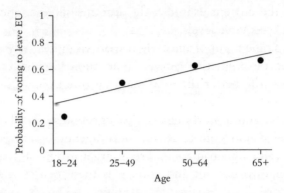

Figure 5.1 Regression model of probability that a person will vote to leave the EU dependent on age. Circles are measurements taken from opinion poll data collected by YouGov in the run-up to the vote as to whether the UK should leave the EU in 2016.[3] Solid line is a model fit, relating age to probability of voting to leave.[4]

What data analytics companies do in order to make predictions is use models fitted to one group of people to infer the preferences of others. Given a person's age, Figure 5.1 can be used to look up their probability of voting to leave the EU. Based on the model, they would infer that a 'typical' 22-year-old would have around a 36 per cent probability of wanting to leave the EU, while a 'typical' 60-year-old would have a 62 per cent leave probability.

Regression models are not a perfect representation of the data. In the survey, only 25 per cent of people aged 18–24 said they were in favour of Brexit (see points in Figure 5.1). So the model slightly overestimated the probability of young people wanting to leave. This type of inconsistency is typical for regression models that attempt to represent a large number of data points (in this case people's ages and voting intentions) as a single equation. I'm bringing up this point as a word of warning, rather than as a serious problem. The inconsistency doesn't mean the model is wrong, it just reflects a general limitation of the regression method. A small inconsistency isn't a major concern – all models are wrong to some degree and in the current example, the amount of 'wrongness' lies within acceptable levels.

The single input of age gives my Brexit model only a small amount of predictive power. The more inputs that are available, the better the prediction becomes. For the Brexit vote, pollsters found that older people with less formal education and a working-class background were likelier to say they would vote to leave.[5] If an agency employed by a 'leave' campaign had to choose a target group of people to encourage to get out and vote, it is these people it should focus on. Pro-remain campaigners would prefer that university students go to the polls.

Political scientists have been using the regression approach for a long time. In one study conducted after the 1987 UK general election researchers measured voters' gender, age, social class and perceptions about inflation, and looked at how these factors affected the probability of preferring the Labour Party to the Conservatives.[6] The researchers then showed that older people and males were more likely to vote Conservative, while working-class people with a perception that inflation was high were more likely to vote Labour. The inputs to the regression model – gender, age, class and inflation perception – were fed into the model and the output was the probability that person voted Labour.

Cambridge Analytica and other modern data analytics companies use more or less the same statistical techniques as were used in the 1980s. The major difference between now and then is the data they have access to. It is possible to feed Facebook likes, answers to online poll questions and data on the purchases we make into regression models. Instead of relying on just age, class and gender to characterise us, Cambridge Analytica claims to use these large data sets to establish an overall view of our personality and political standpoint. In the past, when political scientists studied voters' party preferences, they typically relied on socio-economic background. CA claims to: 'take into account the behavioral conditioning of each individual [voter] to create informed forecasts of future behavior'.[7]

To do large-scale regression on our political personalities, Cambridge Analytica needed a lot of data. In 2014, psychologist Alex Kogan, a researcher at Cambridge University, was collecting data for his scientific studies through an online crowd-sourcing marketplace called Mechanical Turk. Alex described Mechanical Turk to me as 'a big pool of folks who'll do tasks in exchange for cash'. For his scientific study, he was asking them to complete a seemingly inconsequential job: they answered two questions about their income and how long they had been on Facebook, and then clicked a button that gave Alex and his colleagues consent to access their Facebook profile.

The study was a dramatic demonstration of how willing people are to allow researchers access to their own and their friends' Facebook data. It was also, at that time, surprisingly easy for researchers to access data on the social network site. By getting permission from the Mechanical Turk workers, it was also possible to access the location and the likes of their friends. Eighty per cent of people volunteering for Alex's study provided access to their profile and their friends' location data in exchange for $1. The workers had, on average, 353 friends. With just 857 participants, Alex and his co-workers gained access to a total of 287,739 peoples data. This is the power of the social network: collecting data from a small number of people gives researchers access to the data of a vast network of friends.

It was at this point that Alex started talking to representatives for SCL, a group of companies that provide political and military analysis for clients throughout the world. Initially, SCL was interested in Alex helping with questionnaire design. But when the company's representatives realised the power of data collection on Mechanical Turk, the discussion turned to the possibility of accessing vast quantities of Facebook personality data. SCL was poised to set up the political consultancy service, which would later become Cambridge Analytica, to use personality predictions to help its clients win elections. Alex appeared to have exactly the approach to data collection that SCL needed.

Alex admitted to me that he had been naïve. He had never worked with a private company before, having been in academia throughout his undergraduate degree at Berkeley, his PhD in Hong Kong and now his research position at Cambridge. 'I didn't really appreciate how business is done,' he told me.

He and his colleagues considered the ethical aspects and the risks of working with SCL, making sure they separated the data collection from their university research work. They realised that Mechanical Turk didn't have the capacity or the reliability to collect data on the scale required. So they used an established online customer survey service called Qualtrics. Alex told me that, as in their earlier studies, they asked permission to use the respondents' Facebook profiles and followed all the access rules that existed at that time.

What Alex hadn't considered was other people's feelings and perceptions when they heard about the Facebook data collection. 'It is pretty ironic, if you think about it,' he said. 'A lot of what I study *is* emotions and I think if we had thought about whether people might feel weird or icky about us making personality predictions about them, then we would have made a different decision.'

The *Guardian* newspaper later found out that, with financing from SCL, a company set up by Alex collected Facebook data and questionnaire answers for 200,000 US citizens.[8] And that is just the people they surveyed directly. Since the way Facebook's platform operated at the time allowed access to the 'likes' of friends of people who volunteered for the study and who consented to their friends' data being accessed, in total SCL had data for over 30 million people. This was a massive dataset that potentially gave a comprehensive picture of the political personality of many Americans.

CEO of Cambridge Analytica, Alexander Nix, did not appear particularly concerned about people feeling 'icky' about his company predicting their political personalities when he presented his company's research at the Concordia Summit in 2016.[9] His company had just helped presidential candidate Ted Cruz out of obscurity to become a leading candidate in the Republican presidential primaries. Nix talked about how, instead of targeting people on the basis of race, gender or socio-economic background, his company could 'predict the personality of every single adult in the United States of America'. Highly neurotic and conscientious voters could be targeted with the message that the 'second amendment was an insurance policy'. Traditional, agreeable voters might be told about how 'the right to bear arms was important to hand down from father to son'. He claimed that he could use 'hundreds and thousands of individual data points on our target audiences to understand exactly which messages are going to appeal to which audiences' and implied that the methods he had described were being used by the Trump campaign.

The origins of Cambridge Analytica has all the ingredients of a modern conspiracy story. It involves Ted Cruz, Donald Trump, data security, the psychology of personality, Facebook, underpaid Mechanical Turk workers, big data, Cambridge University academics, right-wing populist Steve Bannon who sits on the board, right-wing financier Robert Mercer who is one of its biggest investors, one-time national security advisor Michael Flynn who has acted as consultant, and (in some less reliable versions of the story) Russian-sponsored trolls. I can imagine it as a film with Jesse Eisenberg playing a psychologist who gradually uncovers the true motives of the company he works for: to manipulate our every emotion for political means.

Looked at in this way, it *is* a frightening story. But when I focused on the details of the models used to predict voting patterns, I felt that one important ingredient was missing: the algorithm. I wanted to work out for myself whether Nix's big claims could really hold up to scrutiny.

I don't have access to the data that Alex Kogan collected – I'll come back to what might have happened to that later – but Michal Kosinski and his colleagues have created a tutorial package to allow psychology students to practise creating regression models on an anonymised database of 20,000 Facebook users. I downloaded the package and installed it on my computer. Only 4,744 of the 19,742 US-based Facebook users in the dataset expressed a preference for either Democrats or Republicans.[10] Of these, 31 per cent were Republicans.

Democrats were, at the time the data was collected between 2007 and 2012, over-represented on Facebook. I used the data to fit a regression model with the 50 Facebook dimensions as inputs. The output of the regression model is the probability that a person is a Republican.

After fitting the model to the data, the next step is to test its performance. A good way to test the accuracy of a regression model is to pick two people at random, one Democrat and one Republican, and ask the model to predict which of the pair is the Republican from their Facebook profile. This is an intuitive measure of accuracy. Imagine you met these two people, and you were allowed to ask them a few questions about their tastes and hobbies, after which you had to decide which person supported which political party. How often do you think you would get it right?

The accuracy of a regression model based on Facebook data is very good. In eight out of nine attempts, the regression correctly identifies the political views of the Facebook user. The main group of likes that identify a Democrat were for Barack and Michelle Obama, National Public Radio, TED Talks, Harry Potter, the I Fucking Love Science webpage and liberal current affairs shows like *The Colbert Report* and *The Daily Show*. Republicans like George W. Bush, the Bible, country and western music, and camping.

It isn't too surprising that Democrats like the Obamas and *The Colbert Report* or that many Republicans like George W. Bush and the Bible. So I tried to see if I could break the regression model by taking some of the obvious 'likes' out of the model and performing a new regression. To my surprise, the model still worked with 85 per cent accuracy, only a slight reduction in performance. Now it used combinations of likes to determine political affiliations. For example, someone who liked Lady Gaga, Starbucks and country music was more likely to be a Republican, but a Lady Gaga fan who also liked Alicia Keys and Harry Potter was more likely to be a Democrat. This is where the multiple dimensional understanding gained by using lots of 'likes' produces unexpected and useful results.

This type of information could be very useful to a political party. Instead of Democrats focusing a campaign purely around traditional liberal media, they could focus on getting the vote out among Harry Potter fans. Republicans could target people who drink Starbucks coffee and people who go camping. Lady Gaga fans should be treated with caution by both sides. Although it is difficult to make a direct comparison, the accuracy of a Facebook-based regression model seems to beat traditional methods. For example, in the study of the 1987 UK general election, researchers found that the probability of a middle-class, 65-year-old male voter, who thought inflation was low, preferring the Conservative party over Labour, was around 79 per cent. So a model that assumed these 'typical Tories' were, in fact, sympathetic to the Conservatives would be wrong at least 21 per cent of the time.

So far so good for Alexander Nix and Cambridge Analytica. But before we get carried away, let's look a bit more closely at the limitations.

First of all, there is a fundamental limitation of regression models. Remember, the output of algorithms isn't binary. nor, as we saw in Figure 5.1, is a model a perfect representation of data. We can't expect a model to reveal your political views with 100 per cent certainty. There is no way that Cambridge Analytica, or anyone else for that matter, can look at your Facebook data and draw conclusions with guaranteed accuracy. Unless that is, you happen to be Barack Obama or Theresa May. Otherwise, the best that analysts can do is use a regression model that assigns a probability of you holding a particular view.

While regression models work very well for hardcore Democrats and Republicans – as I established earlier, the accuracy is around 85 per cent – predictions about these voters are not particularly useful in a political campaign. Known party supporters' votes are more or less guaranteed, and they don't need to be targeted. In fact, the regression model I fitted to Facebook data does not reveal anything about the 76 per cent of people who didn't register their political allegiance. While the data shows us that Democrats tend to like Harry Potter, it doesn't necessarily tell us that other Harry Potter fans like the Democrats. This is the classic problem inherent to all statistical analyses; of potentially confusing correlation with causation.

A second limitation relates to the number of 'likes' needed to make predictions. The regression model only works when a person has made more than 50 'likes' and, to make really reliable predictions, a few hundred 'likes' are required. In the Facebook data set, only 18 per cent of users 'liked' more than 50 sites. After this data was collected, Facebook has succeeded in increasing the number of sites its users 'like', precisely so that it can better target advertising. But there are still a lot of people, myself included, who don't 'like' very much on Facebook. I 'like' a grand total of four pages: my own *Soccermatics* page, a local nature reserve, my son's school and European Union research. No matter how good a regression technique is, without data a model can't work.

The third limitation goes to the heart of Nix's idea of targeting our political personalities: can an algorithm reliably identify neurotic or compassionate people from their 'likes'? The data set I used included the results of a personality test that measured the Big Five personality traits. I used this to test whether a regression model could determine which individual, out of a randomly selected pair, was most neurotic. It simply couldn't. I picked two people from the data set at random and looked at their neuroticism scores from the personality test they had performed. I compared these scores to a regression model based on Facebook 'likes'. The personality test and the regression model produced the same rankings for these pairs in only 60 per cent of cases. If I had set scores at random, I would have been correct 50 per cent of the time. The model was only slightly better than random.

The regression model was a bit better at classifying people in terms of their openness: it was correct about two-thirds of the time. But when I performed the same test on extroversion, conscientiousness and agreeableness I got similar results as for neuroticism: the model got it right six times out of 10, compared with the five times out of 10 we would expect if we assigned people at random.

It was at this point I started to discuss my results with Alex Kogan, the Cambridge psychologist who had initially helped CA with its data collection. Initially, he had been reluctant to talk to me, since he had felt unfairly portrayed in the *Guardian* article and a number of online blogs, about Cambridge Analytica.[11] But when I told him my findings about predicting personality from Facebook he started to open up.

Alex had reached similar conclusions to my own. He didn't believe that Cambridge Analytica, or anyone else, could produce an algorithm that effectively classified people's personality. He was working with a combination of computer simulations and Twitter data to show that, although aspects of personality could be measured from our digital footprint, the signal wasn't strong enough to make reliable predictions about us. He was blunt about Alexander Nix. 'Nix is trying to promote [the personality algorithm] because he has a strong financial incentive to tell a story about how Cambridge Analytica have a secret weapon.'

There is an important distinction to be made here between a scientific finding – that a certain set of 'likes' on Facebook is related to the outcome of personality tests – and the implementation of a reliable algorithm based on this finding, creating an equation that correctly predicts what type of person you are. A scientific finding can be true and interesting, but unless the relationship is very strong (which it isn't in the case of personality prediction) it doesn't allow us to make particularly reliable predictions about an individual's behaviour.

One reason that the distinction between scientific results and applied algorithms becomes blurred lies in the way results like this are reported in the media. In January 2015, *Wired* magazine wrote an article titled: 'How Facebook knows you better than your friends do'. The *Telegraph* newspaper in the UK went one step further and ran the headline: 'Facebook knows you better than your [sic] members of your own family'. Not to be outdone, the *New York Times* topped all other media outlets by going with: 'Facebook knows you better than anyone else'.

All of these headlines were followed by a report on the same scientific article. The research – conducted by Wu Youyou, Michal Kosinski and David Stillwell – looked at how well Facebook 'likes' predicted answers to personality questions, but this time, compared a regression model based on likes to answers from a 10-item questionnaire that work colleagues, friends, relatives and partners filled in about the Facebook user. The scientific result, which the newspapers were attempting to capture in their various headlines, was that their statistical model correlated better with the personality test than the 10 answers made by friends and family.

A better correlation implies better prediction, but does it imply that Facebook knows you better than anyone else? Of course it doesn't. I asked Brian Connelly, associate professor at the Department of Management at the University of Toronto, Scarborough, who studies personality in the workplace, what he thought about the study. 'Michal [Kosinski]'s work is interesting and provocative, but I think the media are sensationalising the findings,' he told me. 'A more appropriate headline like, "Preliminary findings suggest that Facebook knows some of you about as well as a close acquaintance (but we're holding out to see whether Facebook can predict your behaviour)", isn't very splashy.' Brian's revised headline sums it up. The science is interesting, but there is no evidence yet that Facebook can determine and target your political personality.

The story of Cambridge Analytica took me deep into a web of blogs and privacy activists' websites. Following these links, I found my way to a YouTube video of a young data scientist, who now works for Cambridge Analytica, presenting a research project he had carried out when working as an intern at the company. He starts his presentation with a reference to the film *Her*, in which the lead character Theodore, played by Joaquin Phoenix, falls in love with his operating system (OS). In the film, the computer forms a deep understanding of Theodore's personality, and the human and the OS fall in love. The young data scientist uses this story to set up his own five-minute presentation: 'Can a computer ever understand us better than a human?'

The data scientist says it can. His presentation takes us, step by step, through the research on online activity and personality. He describes the Big Five personality traits; he outlines how surveys can be replaced by Facebook profiles; he claims that the results of one of his regression models reveals our conscientiousness and neuroticism; he talks about how political messages can be targeted to the individual and then he closes by claiming that 'my model, given your Facebook likes, your age and your gender, can predict how agreeable you are just as well as your spouse'. One day, he says, we might fall in love with a computer that understands us better than our partner.

I start to doubt whether the data scientist in the video truly believes what he is saying. I'm not sure he even expects his audience to believe it. His 'research' is the product of an eight-week training programme at ASI Data Science, a programme for aspiring data scientists. But even if this is just some sort of practice talk, I am deeply disturbed by what I see. This is a young man with the highest level of scientific training: a PhD in theoretical physics from Cambridge University. So I find it difficult to believe that he doesn't have some of the same doubts that I have. 'On what data is this based?', I would like to ask him. 'Have you tested and validated your model over time? What about the fact that likes measure neuroticism only slightly better than random?'

It appears that the ASI fellowship has encouraged him to set these doubts aside when presenting his research project. The result for him personally was a job offer from Cambridge Analytica, which he duly accepted.

I don't know this young man, but I do know lots of others like him. I work with them and I train them as PhD students, as master's students and as undergraduates. What I felt, as I watched the video, was a deep sense of failure on my part. There is a demand from companies like Cambridge Analytica that the universities provide them with this type of ambitious young person: people who can both do research and present their results in a way that is easy to understand.

We live in an exciting time, where we can use data to help us make better decisions and keep people informed about the issues that are important to them. But with this power comes the responsibility to carefully explain what we can and can't do. When my colleagues and I train researchers we tell them about their powers, but we often forget to tell them about their responsibilities. It seems we have left this important job in the hands of industry consultants who are teaching data scientists how to spin their research to the greatest possible effect.

Either the young man in the video is outnumbered – unable to see the limitations of the method he is presenting – or he is trying to outnumber his audience – by neglecting to mention these limitations. Instead of claiming that 'my' algorithm knows you as well as your partner, a careful scientist would say 'for individuals who use Facebook a lot, one study by Michal Kosinski and co-workers showed that their "likes" can be used to predict their personality score, but what this implies about personality-based marketing remains unclear'. Unfortunately, like Brian Connelly's headline, this latter statement doesn't have the same ring to it. Nor is it a point that his future employers want to have as part of a five-minute showcase of their methodology. Careful science doesn't sell a political consultancy service.

A few months into the Trump presidency, Cambridge Analytica removed the reference to the Big Five personality model from their webpage. I heard from a reliable source that Facebook had told them to delete all of the data it had collected about its user 'likes' *before* they started working with the Trump campaign. CA claims it complied with Facebook's demands. It is unlikely that they even tried to carry out the targeting described by Alexander Nix at the Concordia Summit, in connection with the Trump campaign. CA has since stated that none of the Facebook data it received from Alex Kogan was used in the services it provided to the Trump campaign, nor was personality targeted advertising used in that campaign generally.

In January 2017, David Carroll, associate professor at Parsons School of Design in New York City, made a data protection request to Cambridge Analytica. CA replied with a list of information they held about him. They had stored his age, gender and place of residence. They had a spreadsheet showing the district he had voted in, including a column indicating that he had voted in the Democrat primary. They had used this data to rank the importance they thought he would give to various issues, such as the environment, healthcare and the national debt. They noted that he was a 'very unlikely Republican' with a 'very high' propensity to turn out to vote in the election. After all the hype, Cambridge Analytica was using old-school regression methods based on age and place of residence, to predict David's vote. The data held and the methods used were a long way from being the personally targeted political adverts that Alexander Nix had boasted about.

The Cambridge Analytica story* is in my view primarily one about hyperbole. It is a story about a company seemingly exaggerating what they can do with data. Alexander Nix has himself admitted to 'speaking with a certain amount of hyperbole' about what CA does. But it is just one case study. With everyone from Facebook and Spotify to travel agencies and sports consultants purporting to offer algorithms that rank us and explain our behaviour, I needed to find out more about the accuracy of these algorithms. How well do these algorithms really know us? And are they making other, even more dangerous, mistakes?

---

* The hyperbole created by CA around their approach exploded on a massive scale as this book went to press. For more information, see endnote.[12]

CHAPTER SIX

Impossibly Unbiased

Dissecting personality algorithms changed my perspective, but not in the way I expected. I was less concerned about the algorithms making dangerously accurate predictions about us and more worried about how they were being marketed. The conclusions I'd drawn about Cambridge Analytica were similar to those I had drawn, more tentatively, after reading the Banksy article. In the latter case, the researchers already needed to know who Banksy was to track him down. Algorithms are useful for organising data in a political campaign or in a criminal investigation, but it wasn't simply a case of pressing a button and finding a graffiti artist or a list of neurotic Republicans.

Algorithms are regularly marketed as providing insight into who we are as people and as being able to predict how we might behave in the future. They are used to decide whether or not we get a job, if we are given a loan, or if we should be sent to jail. I felt I needed to know more about what was going on inside these algorithms and the types of mistakes they might make.

In the US, the COMPAS algorithm is used in some states to help make risk assessments of criminal defendants, usually when they are seeking parole. Some media reports have described COMPAS as a black box, implying that it is difficult or impossible to know what is going on inside. I contacted Tim Brennan, creator of the COMPAS algorithm and director of the company Northpointe that supplies it, and asked if he would be willing to explain how the model works. After a few emails back and forth, he sent me the internal reports that explained how the scores were created.[1] When I later interviewed him, he was reasonably open about the model, pointing me towards the specific equations needed to understand it.

Tim's model uses a combination of the defendant's criminal record, age at first arrest and age now, and education level, along with answers to an hour-long questionnaire-based interview, to predict whether they will reoffend. All of these measurements are then used to fit a statistical model, based on previous defendants. People with a history of non-compliance or violence are more likely to reoffend, as are people with lower levels of education or who use drugs.[2] People with financial problems or who move house a lot are *not* more likely to reoffend. It is these patterns in the population as a whole that the model uses to make predictions.

The techniques employed inside COMPAS are similar to those I had seen so far: the data is first rotated and reduced using PCA, and then a regression model is used to predict reoffending based on historical results. I wouldn't claim that it was easy to follow the details as an outsider. The technical reports were hundreds of pages long, but the model is fully documented and Tim pointed me to the most relevant parts. After dealing with Cambridge Analytica, I was impressed with Northpointe's openness.

The fact that an algorithm's creators are open about the details does not mean that their algorithm gets things right. In 2015, an article by Julia Angwin alleged that the algorithm was biased against African Americans.[3] ProPublica used the only failsafe way to determine whether or not an algorithm is unbiased: to look at the quality of the predictions it makes. COMPAS assigns a score between 1 and 10 related to the probability of the defendant being arrested for an offence in the future. Julia and her colleagues' results were clear. They found that 45 per cent of the black defendants who scored as higher risk, and thus more likely to be imprisoned, had been placed in too high a risk category. This compared with an error level of 23 per cent for white defendants. Black defendants who didn't go on to commit a crime were more likely to be wrongly classified as high risk than white defendants.

When Julia and her colleagues published their article, Tim and Northpointe were quick to respond. They wrote a research report arguing that the ProPublica analysis was wrong.[4] They argued that COMPAS held the same standards as other tried and tested algorithms. They claimed that Julia and her colleagues had misunderstood what it means for an algorithm to make an error and that their algorithm was 'well-calibrated' for white and black defendants.

The debate between Northpointe and ProPublica made me realise just how complicated the issue of bias was. These were smart people, and their exchange of words covered nearly 100 pages of arguments and counter-arguments. It was supplemented by computer code and additional statistical analyses. The two interlocutors' debate was then commented on by bloggers, mathematicians and journalists, all of whom added their own perspective to the problem. Defining bias was a difficult mathematical problem, which I needed to look at in detail if I was going to understand it.

So I downloaded the data ProPublica had collected and got to work.

To understand ProPublica's argument and Northpointe's counter-argument, I started by redrawing the table showing how the COMPAS algorithm classified white and black defendants and whether or not they were arrested for a further offence. These data collected by ProPublica from Broward County, Florida, are shown in Table 6.1. The columns are the number of people categorised as higher risk and lower risk by the COMPAS algorithm. The rows are the number of people who did and didn't go on to reoffend.

You should take a minute now to look at this table and think about whether or not you believe the algorithm is biased. First of all, compare how many black and white people are classified as higher risk. There were a total of 2,174 black people classified as high risk out of 3,615 total black defendants. So the probability a black defendant is classified as higher risk is 2,174/3,615 = 60 per cent. If you make a similar calculation for white defendants we find that the probability of being classified as higher risk is only 34.8 per cent. So black people are more likely to be classified as higher risk than white people.

Table 6.1 Breakdown of the risk assessments on the basis of the COMPAS algorithm (columns) and whether or not the person went on to commit an offence in the two years after the assessment (rows). Details of how 'higher' and 'lower' risks were defined and other details can be found in ProPublica's analysis.[5]

Black defendants	Higher risk	Lower risk	Total	White defendants	Higher risk	Lower risk	Total
Re-offended	1,369	532	1,901	Re-offended	505	461	966
Didn't reoffend	805	990	1,714	Didn't reoffend	349	1,139	1,488
Total	2,174	1,522	3,615	Total	854	1,600	2,454

In itself, this difference does not imply that the algorithm is biased, because the proportion of reoffenders varies between black and white defendants: 52.9 per cent of black defendants were arrested for another offence within two years, while 37.9 per cent of white defendants were arrested for another offence. This difference in the overall proportions of those classified as higher or lower risk was not the basis of ProPublica's critique of the algorithm. Julia and her colleagues recognised that there was a higher rate of recidivism for black defendants than white defendants, then asked what types of mistakes the algorithm made.

When evaluating algorithms, it is often useful to think in terms of 'false positives' and 'false negatives'. For the COMPAS algorithm, a false positive is when a person who is not going to commit a crime in the future is assigned as high risk, *i.e.* the model prediction was positive but wrong (false). The false positive fraction is the number of higher risk non-reoffenders divided by the total number of people who didn't reoffend. For black defendants, this is 805 divided by 1,714, which is 46.9 per cent. For white defendants it is 23.5 per cent. Black defendants had a much higher proportion of false positives than white defendants.

If the police have detained you, and a judge is using an algorithm to help assess you, the worst thing that can happen is that you get a false positive result. A true positive is fair: the algorithm predicted that you were a risk and you were. But a false positive could mean that you are denied parole or given a longer sentence than you deserve. And this was happening more often to black defendants than to white ones. Nearly half of the black defendants who did not go on to reoffend had been labelled high risk.

In the other direction, white defendants experience more false negatives, where the algorithm says that a person is low risk but they commit a further crime. For white defendants, the false negative fraction is $461/966 = 47.7$ per cent and for black defendants it is just $532/1,901 = 28.0$ per cent. For society, a high false negative fraction is a problem; it means people who should have been detained have been let back out into society and have committed crimes. Nearly half the white people who reoffended had been labelled as lower risk.

Seen in terms of these false positive and false negative rates, the algorithm looks very bad indeed. COMPAS could potentially be used to send black people to jail for unnecessarily long periods, while letting off white people who go on to commit crimes.

Replying to this accusation, Northpointe argued that its algorithm should instead be judged in terms of whether or not its predictions were equally good for black and white people. And they are. If we look at the first column of Table 6.1, we see that 1,369 of black defendants classified as high risk, out of a total of 2,174, went on to reoffend. That is 63.0 per cent. For white defendants, 505 out of 854, or 59.1 per cent, of people classified as high risk went on to reoffend. These proportions are similar, so the algorithm is properly calibrated for both white and black defendants. If a judge is given a risk score for a particular person then, irrespective of race, this score reflects probability of reoffending.

These two different approaches to measuring bias produce contradictory results. Julia and her ProPublica colleagues' argument about false positives and false negatives is powerful, but Tim and his Northpointe colleagues' counter-argument about algorithm calibration is solid. Given the same table of data, two separate groups of professional statisticians had drawn opposite conclusions. Neither of them had made a mistake in their calculations. Which of them was correct?

It took two Stanford PhD students, Sam Corbett-Davies and Emma Pierson, working together with two professors, Avi Feller and Sharad Goel, to solve the puzzle.[6] They confirmed Northpointe's claim that Table 6.1 showed the COMPAS algorithm gave equally good predictions, independent of race. Then, as mathematicians like to do, they pointed out a more general problem. They showed that if an algorithm is equally reliable for two groups, and one group is more likely to reoffend than another, then it is impossible for the false positive rates to be the same for both groups. If black defendants reoffend more frequently, then they have a larger probability of being incorrectly placed in a higher-risk category. Any other result would mean that the algorithm was not calibrated equally for both races, in the sense that it would have to make different evaluations for white and black defendants.

To better understand this point, let's conduct a thought experiment. Imagine I want to create an online job advert on Facebook to hire a computer programmer for my research group. This is a simple thing to do. I create a job advert as a post on my research group's Facebook page. I then click on the 'boost post' button that makes it easy to target my advert.[7] By using the 'create audience' feature, I can find people who are interested in dogs, people who are war veterans, console gamers or motorcycle owners. I can find hobbies such as acting, dance and guitar playing.

Facebook doesn't have a box that allows me to separately target men or women, nor do I think it should. But I am aware that, due to different educational choices at school and university, there happen to be more men interested in the programming job than women. Let's assume, for sake of argument, that out of a population of 1,000 women, 125 are interested in the job as a programmer, while out of a population of 1,000 men, 250 are interested.

When I make my advert, I decide to click a few boxes that I think appeal to computer programmers: 'role-playing games', 'science fiction movies' and 'manga comics'. That should do it. Remembering my own time as a computer science student, I know that lots of programmers love these activities. I'll draw in some good applicants that way, and won't have to waste money advertising to people who are not interested in the job.

I put up the advert and wait.

After a day, Facebook has shown my advert to 500 people: 100 women and 400 men.

When I tell you the result, you are appalled. 'Your advertising campaign is biased,' you tell me. 'Role-playing games? Sci-fi? Those boxes you clicked don't just appeal to computer nerds, they also typically appeal to more men than to women. Your algorithm is unfair!'

'But look,' I say, 'I've done the stats. My algorithm is unbiased.' I get out Table 6.2 and show it to you. I mathsplain it in my most patronising, superior tone: 'Of the 100 women to whom the algorithm showed the advert, 50 of them were interested in the job and would have gone on to apply. Of the 400 men who saw it, 200 were interested. So the advert was equally relevant to the men who saw it as it was to the women who saw it.'[8]

'But four times as many men saw it!', you cry, frustrated with my perverse numerology, 'and you knew from the start that at least half as many women as men could be interested

Table 6.2 Breakdown of the men and women shown an advert in my (thought experiment) Facebook campaign.

Women	Shown advert	Not shown	Total	Men	Shown advert	Not shown	Total
Interested in job	50	75	125	Interested in job	200	50	250
Not interested	50	825	875	Not interested	200	550	750
	100	900	1,000		400	600	1,000

in the job. You are exaggerating society's inbuilt bias even further.'

You are right, of course. It isn't fair that I created an advert that was shown to four times as many men as women. I have used the same logic Northpointe used to justify its algorithm. I used the calibration definition of unbiased: the proportion of individuals correctly predicted to be interested in the job is the same for both groups. This is what Tim Brennan argued when he showed that the criminal outcome for defendants is the same for both groups. My advertising campaign has placed a premium on eliminating calibration bias.

You are patient with me. You fill in a few extra boxes in the Facebook ads algorithm. We run it together this time and look at the results (shown in Table 6.3). Now the algorithm shows the advert to 100 women with a potential interest in applying for the job and 200 men who might be interested. The balance 100 to 200 reflects the underlying proportion of the total men and women interested in the job (125 to 250). The proportion of false negatives (one in five) is the same for men and women.

Here is the catch that, even if I have now accepted your way of doing things, I can't help pointing out. Only one-third of the women who saw the advert were interested in the job, while half the men who saw the ad were interested. Moreover, if we consider the individuals who weren't shown the advert, we could be said to have discriminated against males. One in 11 of the men who didn't see the advert were interested in the

Table 6.3 Revised breakdown of the men and women shown an advert for our revised (thought experiment) Facebook campaign.

Women	Shown advert	Not shown	Total	Men	Shown advert	Not shown	Total
Interested in job	100	25	125	Interested in job	200	50	250
Not interested	200	675	875	Not interested	200	550	750
	300	700	1,000		400	600	1,000

job, while only one in 27 of the women who didn't see the advert were interested. Our new algorithm has a calibration bias that favours females.

Unfairness is like those whack-a-mole games at the fairground where the mole keeps popping up in different places. You hammer it down in one place and another one comes out somewhere else. Try it yourself. Make two empty two-by-two tables and try placing 1,000 women (of whom 125 are interested in the job) and 1,000 men (of whom 250 are interested in the job) between the four boxes in a way that is totally unbiased. You can't do it. It is impossible to have both calibration between groups and equal rates of false positives and false negatives for men and women. There is always some group who is discriminated against.

The beauty of mathematics is that we can prove general results. This is exactly what Cornell computer scientists Jon Kleinberg and Manish Raghavan, working with Harvard economist Sendhil Mullainathan, did for pairs of two-by-two frequency tables, similar to those in Tables 6.2 and 6.3. I chose specific combinations of numbers in my example, but Jon, Manish and Sendhil showed, in general, that it is impossible to eliminate both calibration bias and have identical false positive and false negative rates for two groups.[9] This result applies irrespective of the numbers we put into the table, with only one notable exception: if the underlying features of the groups are exactly the same. So only if reoffending rates were identical for black and white defendants in Broward County, Florida, or if as many women study computer programming as men, could we hope to create totally unbiased algorithms. When the world we live in isn't equal in every way possible, then we can't expect our algorithms to be completely fair.

There isn't an equation for fairness. Fairness is something human. It is something we feel. My feeling is that you were correct when you changed my advertising algorithm. I instinctively prefer Table 6.3 to Table 6.2 for an advertising campaign. When trying to find the best person for a job opening, it feels wrong to make an advert that produces disproportionally more male applicants than female applicants. And to me, it feels right that we should invest time in building algorithms that are better at finding qualified women for a programming job, even if that means that our algorithm is not equally good at finding men.

I also felt that Tim Brennan and the other creators of the COMPAS algorithm were wrong to emphasise the elimination of calibration bias in their score's predictions. If Northpointe could create an algorithm that was better at identifying whether black people were higher risk or not then, even if the algorithm didn't work as effectively for white people, I wouldn't consider it racial discrimination. The algorithm would be addressing an important problem in society.

During my investigations of the ProPublica dataset, I found one interesting lead that might help create an algorithm with a better false positive rate. Very little of the debate around the COMPAS algorithm addressed the main reason that black defendants in Broward County reoffend more often than white defendants. It is simple, really. Black defendants are typically younger when arrested.[10] And young people, in general, are more likely to reoffend.[11] So if Northpointe could find a better way of identifying younger people that, despite being arrested for one crime, are less likely to offend in the future, then most of us would agree that this would be a good thing. Such a method would inadvertently create a calibration bias between white and black people; because black defendants are younger than white defendants, a model that improved performance for young people would perform better for black defendants on average.

The question I wanted to ask Tim was whether calibrating his algorithm was really that important, when he could instead be thinking of ways of keeping young black men and women – who have possibly made just one stupid mistake – out of jail?

A few days after I finished my analysis, I managed to set up an interview with Tim and asked him what he thought about my idea. He listened patiently and agreed that age, together with criminal record and drug use, was one of the most important factors in predicting reoffending. But he emphasised that in the US there are 'constitutional requirements that demand an equity between races'. According to a Supreme Court ruling, models must be equally accurate for all groups (in the sense of having no calibration bias) unless there is a very strong public concern with the particular issue. So he and his colleagues were constantly 'walking a tightrope' between improving accuracy and following these requirements.

Tim is certain that the statistical tests prove his model is unbiased and cited several independent reports that back up this claim.[12] He told me that the ProPublica report had made people think more critically, but had also been a distraction from the more important debate about using rigorous statistical methods in sentencing. 'If the accuracy levels of sentencing judges is taken into account, then risk assessments are far ahead of human judgment, especially with regards to the false positive errors that can [disproportionally] affect black defendants' he told me.

Before ProPublica's study of algorithms in criminal sentencing, Jen Skeem, professor at the Goldman School of Public Policy at Berkeley, had conducted a comprehensive assessment of a sentencing algorithm called PCRA. She concluded that it was equally well calibrated for black and white defendants, and should not be labelled as biased. 'These issues around bias are not new,' she told me, 'it's just become popular at the moment to rage against the "biased algorithm".'

Jen told me that the most important question is typically overlooked: 'How does "bias" compare with existing practices?' It is this question that she is researching now.

I began to realise how difficult it was to pick the good guys and the bad guys in this story. My arguments for removing bias from algorithms were based on my personal experience and my values. Even if my point of view happens to be right in a moral sense, it wasn't correct in a mathematical proof type of way. All that maths had shown me was that there is no equation for fairness. It was clear that Jen and Tim were just as passionate about using algorithms for good as

Julia Angwin, Cathy O'Neil and Amit Datta, who we met in Chapter 2. All of them were trying to do the right thing.

Whenever we turn to mathematics to find the right thing to do, it gives us the same answer: fairness doesn't come from logic alone. There are many other examples of problems of fairness escaping definition to be found in the history of mathematics. Kenneth Arrow's 'impossibility theorem' tells us there is no system for choosing between three political candidates under which all voters' preferences are fairly represented.[13] Peyton Young's book *Equity*, which uses mathematical game theory to treat the subject matter, is by the author's own admission 'a stock of examples that illustrate why equity cannot be reduced to simple, all-embracing solutions'.[14] And Cynthia Dwork and her colleagues' 2012 work 'Fairness through Awareness', resorts to looking at how we can best balance affirmative action for groups with fairness to the individual.[15] Like in Jon Kleinberg and his colleagues' work on bias, when these authors did the maths they found paradoxes instead of rational certainty.

I thought back to the motto, once so proudly stated by Googlers: 'Don't be evil.' It isn't used as frequently at the company now. Had Google abandoned its axiom after one of its mathematicians discovered that there was no equation that could allow them to avoid evil with certainty?

We can try our best, but we can never be truly sure whether or not we are doing the right thing.

CHAPTER SEVEN

The Data Alchemists

Many of the researchers and activists I had talked to up to this point took one thing for granted: algorithms are smart and rapidly getting smarter. Algorithms think in hundreds of dimensions, processing vast quantities of data and learning about our behaviour.

These views came just as often from those with a more utopian vision, like COMPAS creator Tim Brennan who saw a future where algorithms help us with critical decisions, as those with a more dystopian view, like the people blogging angrily about Cambridge Analytica. Both sides believed that computers were either currently outperforming us or would soon outperform us in a large range of tasks.

The impression that we are experiencing a massive change in what algorithms can achieve was reflected in the media. Reporting around the COMPAS algorithm, on Cambridge Analytica and on the power of targeted advertising on Google and Facebook were full of references to the potential dangers of AI.

What I had found so far gave a different picture. When I looked more closely at Cambridge Analytica and political personalities, I'd found fundamental limitations of algorithm accuracy. These limitations were consistent with my own experience of modelling human behaviour. I have worked in applied mathematics for over 20 years. I have used regression models, neural networks, machine learning, principal component analysis and many of the other tools that were gaining prominence in the media. And, during that time, I have come to realise that, when it comes to understanding the world around us, mathematical models don't usually beat humans.

My point of view may sound surprising since I work in the business of using mathematics to predict the world. At the same time as writing this book, I run a company that uses models to understand and predict the outcome of football matches. And I lead an academic research group that uses mathematics to explain the collective behaviour of humans, ants, fish, birds and mammals. I am deeply invested in the idea that models are useful. So, in my position, it probably doesn't pay to cast too many doubts on the usefulness of mathematics.

It is important, though, that I am honest. In my work on football, I meet scouts and analysts at leading clubs. What strikes me when I tell them a statistic about a player's chance creation or game contribution is their intuitive understanding of the underlying reasons for the numbers. I'll say something like 'player X played 34 per cent more dangerous passes than player Y in the same position.'

The scout will say, 'OK, let's look at defensive contribution. There you go ... higher for player Y. The manager is telling him to defend in this situation thus reducing his opportunity to create chances.' While computers are very good at collecting large numbers of statistical measures, humans are very good at discerning the underlying reasons for these measures.

One of my football number-cruncher colleagues, Garry Gelade, recently set about deconstructing a central model from football analytics known as 'expected goals'. The statistical concept behind expected goals is solid. Data is collected on every shot at goal taken in top-tier football: the position in or around the box the shot was made from; whether it was a header or made with the foot; whether it came from a counter-attack or a slower build up; the level of defensive pressure at the time the shot was taken and so on. This data is then used to assign the expected goal value to every shot. Shots taken centrally, within the penalty area and where the striker can see the face of the goal, are given a higher expected goal value. Shots taken at oblique angles or from outside the box are given lower expected goal values. Every shot a team takes is automatically assigned a value between zero (no chance of scoring) to one (guaranteed goal).[1]

Expected goals are useful because they allow us to evaluate how well a team has performed in low-scoring football matches. A match might end 0–0, but the team that created a lot of chances will have a higher expected goals total. These numbers have predictive power: teams with more expected goals in their previous matches tend to score more real goals in their subsequent ones.

When Garry performed his analysis, during summer 2017, expected goals were taking off in the mainstream media. Sky Sports and the BBC were showing the expected goals stats of the English Premier League's summer signings; the *Guardian*, the *Telegraph* and *The Times* all carried articles explaining the concept and in the US the statistic was widely displayed on the Major League Soccer (MLS) and National Women's Soccer League (NWSL) homepages. Expected goals were increasingly accepted as an 'objective' way of measuring how well a team had performed.

Garry compared expected goals to another, more human, way of assessing the quality of a goal chance in football. The sports performance company Opta collects a metric it calls 'big chance'. These are measured by a trained human operator, who watches the match and looks carefully at every shot. If the operator thinks the shot had a good chance of being a goal, then they label it 'big chance'. If they think it wasn't much of a chance they label it as 'not a big chance'. Comparing 'big chance' and 'expected goals' allowed Garry to compare the ability of humans and computers to assess the quality of goal chances.[2]

We can assess 'big chance' accuracy in two ways. Firstly, we can look at the proportion of shots that weren't a goal but were assigned as a 'big chance' by the operator. This value is the false positive fraction, a concept we came across in Chapter 6. False positives are the proportion of misses that the operator labelled as a 'big chance'. Secondly, we can look at the proportion of goals that were labelled as 'big chances'. This is the true positive fraction, where the operator made a correct prediction. For 'big chances', misses were incorrectly predicted in seven per cent of cases (false positive) and goals were correctly predicted in 53 per cent of cases (true positive).

Garry found that the expected goals model was unable to reproduce this same level of accuracy. Depending on how he tuned the model, it either made more false positives or fewer true positives, but it couldn't match the predictive power of big chances. Expected goals models use a lot of data, but they don't (yet) beat humans. It might initially sound impressive that we have an algorithm for measuring performance in football, but the method doesn't outperform an educated football fan (operators are typically recruited from football enthusiasts) making a note each time a team generates a goal-scoring chance.

Garry told me that he carried out his analysis after reading an article describing expected goals as the 'perfect' model of football. Although such hype might have potential short-term benefits for his business, in the long term it can damage the reputation of statistical analysis in football. Garry has worked as a consultant for a range of clubs, including Chelsea, Paris Saint-Germain and Real Madrid. He believes that for now at least, rather than replacing humans, models can help human decision making. He gave me an example of how technology could be used to observe goalkeepers during matches and train their positioning and movement. The approach is practical and down to earth. In each aspect of the game, there are ways models can be used, but there is no 'perfect' model of football.

Glenn McDonald, who works for music-streaming service Spotify, is another data expert with a down-to-earth attitude to his job. With several competing services available, like TIDAL and Apple Music, Spotify's aim is to have an edge over its competitors in terms of suggestions for new music and creating interesting playlists. Spotify achieves this goal by picking up on our listening patterns. Every recommendation it gives, from 'song radio' to the 'just for you' playlists, uses a musical genre system developed by Glenn and his colleagues.

Spotify's genre system places all songs as a point in 13 dimensions, grouping together those close-by points as genres. The dimensions include objective musical properties such as 'loudness' and 'beats per minute', as well as more subjective emotional properties, such as 'energy', 'valence' (sadness) and 'danceability'. These latter, subjective measurements are established through listening sessions, where human subjects listen to pairs of songs and state which of them they think is saddest or most danceable. The algorithm learns the difference and classifies other songs appropriately.

Glenn has created an interactive visualisation called 'Every Noise at Once', which places all 1,536 of Spotify's musical genres out in a two-dimensional cloud. From 'deep opera' at the bottom to 're:techno' at the top; from 'Viking metal' at the leftmost point to 'African percussion' on the right, every conceivable form of music is placed out with similar genres placed close together. It is a remarkable technical achievement, giving a very concise way of viewing a very large part of the world's musical heritage.

I first spoke to Glenn for an article I was asked to write for the *Economist 1843* magazine. Before the interview, I was a bit nervous about revealing my own opinion of Spotify's suggestions. I had used the 'discover weekly' service now and again to find new music, but was often frustrated. I tend to like melancholy songs, but when I listened to the tunes suggested by Spotify they didn't have the same emotional effect as my own sad favourites. In fact, the suggested songs tended to be quite boring. Many Spotify users complain of the same problem: the songs it recommends are watered-down versions of their true favourites.

When I told Glenn that I often found myself flipping through song after song without fastening to any of the recommendations, I expected him to be slightly disappointed. But he was happy to admit his algorithm's limitations. 'We can't expect to capture how you personally attach to a song,' he told me.

Glenn explained that Spotify playlists work best for music at parties. 'It's a collective thing,' he told me. 'We do very well at generating playlists for social occasions and the number of skipped songs is low. But if we are trying to suggest a new song to you as an individual, then we're satisfied if you like every tenth recommendation.' He is right. When my wife and I have friends over we often put on a generic Spotify playlist. It avoids a lot of arguments about music choice and we often enjoy its suggestions.

Glenn told me that the process of making recommendations is far from a pure science, 'half of my job is trying to work out which computer-generated responses make sense'. When Glenn chose his job title, he asked to be called 'data alchemist' instead of 'data scientist'. He sees his job not as searching for abstract truths about musical styles, but as providing classifications that make sense to people. This process requires humans and computers to work together.

Given the vast scope of 'Every Noise at Once', Glenn's modesty resonated strongly with me. Like many of the data scientists I had spoken to, he saw his job as navigating a very high-dimensional space. But he was the first person I had talked to who openly acknowledged the deeply personal and unknowable dimensions of our minds. He talked about the feeling we have when we hear the song we first fell in love to, or the song we played when we drove a car for the first time. He talked about songs that made us realise something about our own lives and songs that had changed our attitude to homosexuality or racism. These were dimensions that Glenn admitted that he was unable to explain.

The concept of data alchemy perfectly captures how modern-day digital marketing operates. Shortly after I spoke to Glenn, I talked to Johan Ydring, head of brand and performance at TUI Nordic. The TUI group encompasses thousands of travel agents and online portals; runs hundreds of aircraft and hotels; and deals with 20 million customers. Johan's job is to make sure the company best utilises all the data it collects about its customers combined with the data it takes in from Facebook and other social media sites.

Johan described his operations as 'pretending to be smart'. His team come up with four or five approaches to marketing specific target groups, and try them out. If an approach seems to be working they try it on a larger group.

Often the simplest ideas are best. If a customer has booked a holiday twice in a row to Spain, then Johan's team makes sure a Facebook advert appears in their feed just before the time of the year they would typically book their next summer break. The advert might suggest Portugal, a place where the customer has never been. This can have a spooky effect on the user, who might feel that Facebook has read their mind. Only the other day they were talking to a friend about the Algarve and now it's there on their screen. In fact, they are experiencing a simple statistical trick. The data alchemists have worked out the time of year when people typically book their holidays, and found a connection between trips to Spain and Portugal.

Most of us have experienced the feeling that Facebook or Google have read our minds. In the evening before he eats his supper, my son is bombarded with YouTube adverts for his favourite bread, the least healthy one available. Recently my wife bought a chocolate brand for the first time in the local shop, and suddenly the exact same brand started to appear as an advert in her Facebook feed.

After experiencing targeted advertising, I often hear my family and friends speculate about how the Internet is watching them. They begin to wonder if WhatsApp might be selling their private messages, or if their iPhone might be recording their conversations.

Conspiracy theories about companies using private messages are unlikely to hold. The more plausible explanation is that data alchemists are finding statistical relationships in our behaviour that help them target us: kids who watch *Minecraft* and *Overwatch* videos eat sandwiches in the evening. My wife might not have noticed that she had already been shown an advert for that chocolate brand on Facebook.

The other major source of 'spooky' adverts is retargeting: we simply forgot that we searched for a trip to the Algarve, but your web browser has remembered and fed this information to TUI, who are now offering you a room in their finest hotel.

We are subjected to such vast quantities of advertising and spend such long periods of time staring at our phones and screens, that now and again adverts appear to have read our minds.

It isn't really the algorithm that is clever. The intelligence comes from the alchemists, who are putting the data together with their own understanding of their customers. Johan and his colleagues' approach is smart because it gets results, sometimes increasing sales tenfold, but it doesn't follow a well-defined scientific methodology. Their approach lacks rigour. Johan told me that even if they had a very clever data scientist working for 10 years on detailed models of his customers, he isn't convinced that it would be worth the investment. 'Our approach needs to work on large numbers of people and the data isn't reliable enough to target specialised small groups,' he told me.

Talking to Garry, Glenn and Johan, I could see that algorithms for classifying people still have a long way to go. As a rule of thumb, algorithms simply don't make as accurate predictions about our behaviour as other people do. Algorithms perform best when they are used by people who understand their limitations.

It was just as I was reaching this conclusion that I found out something else about the COMPAS algorithm. Something I could never have worked out for myself.

While I had been busy writing this book, Julia Dressel, an undergraduate student in computer science at Dartmouth College in New Hampshire, had published a remarkable honours thesis. She had also looked at the model of reoffending, but from another perspective. She wanted to see how well the algorithm performed in comparison with humans.

To compare humans and algorithms, Julia relied on the data obtained by ProPublica on offenders in Broward County, Florida. She used offender sex, age, race, previous misdemeanours and crimes, as well as a description of the charge made, to construct standardised description paragraphs about the offenders. These took the following form:

The defendant is a [RACE] [SEX] aged [AGE]. They have been charged with: [CRIME CHARGE DESCRIPTION]. This crime is classified as a [CRIME DEGREE]. They have been convicted of [PRIORS COUNT] prior crimes.

They have [JUV FEL COUNT] juvenile felony charges and [JUV MISD COUNT] juvenile misdemeanour charges on their record.

Each of the variables (race, sex, etc.) were inserted into the description directly from the Broward County database. This was all the information provided in the description. No psychological test of the offender was conducted. No detailed statistical analysis of previous offences was performed. No interviews were carried out. Julia's question was whether or not, from this brief description, people with no legal training could predict reoffending.

To test her hypothesis, Julia turned to the same tool Alex Kogan had started his work with: Mechanical Turk. She offered Mechanical Turk workers, all of whom were based in the USA, $1 to evaluate 50 different defendant descriptions. After seeing each description they were asked, 'Do you think this person will commit another crime within two years?', to which they answered either 'yes' or 'no'. The workers were told beforehand that they would receive a bonus of $5 if they answered correctly in at least 65 per cent of cases. This provided a bit of extra motivation for them to make good predictions.

Nearly half of the Mechanical Turk workers answered a sufficient number correctly and won the prize. On average, the workers were correct in 63 per cent of cases, so slightly less than half of them won the extra $5.

More importantly, the performance was not significantly worse than the performance of the algorithm.[3] These two methods, human and algorithm, were indistinguishable.

This is a sobering result for advocates of using algorithms in sentencing. Despite all the complexity of the algorithm – the comprehensive data collection; the long interviews with offenders; the use of principal component analysis and regression models; the writing of 150-page operational manuals and all the time taken to train judges in using the algorithm – it produces results that are no better than those achieved by a bunch of people randomly recruited from the Internet. Whoever these people happen to be, we do know one thing for sure about them: they have nothing better to do with their time than to earn $1 playing a criminal-identity guessing game. Amateurs beat the algorithm.

Julia told me that she was motivated to conduct the study because she is frightened by the many ways that technology reinforces oppression. 'Humans are quick to assume that technology is objective and fair, so the cases where technology isn't fair are the most dangerous,' she told me.

The reports about the racial bias of the COMPAS algorithm had initially motivated Julia's research project. She wanted to find out whether humans showed the same bias as the algorithm. In this respect, her results supported Tim Brennan's argument that his algorithm was fair. The Mechanical Turk workers made the same sentencing decisions when they were presented with the race of the defendant as when they were not shown the race, and these decisions matched the COMPAS algorithm performance in almost every respect. So COMPAS is not any more or less racist than Mechanical Turkers.

No, COMPAS is not more racist than we are, but it isn't particularly effective either. Julia summarised her main findings for me very succinctly: 'What I found was that a major commercial software that is widely used to predict recidivism is no more accurate or fair than the predictions of people with little to no criminal justice expertise who responded to an online survey.' I agreed. My own work with the data showed that a model based on just age and number of previous convictions had an accuracy level similar to COMPAS.[4] These were presumably the factors relied upon by Mechanical Turk workers when they made their judgments.

I can't systematically test all algorithms that are used to measure human behaviour and personality. I simply don't have time. But for those models I had investigated in more detail – goals in football, musical tastes, prediction of criminality and political personality – I found the same result: algorithms, at best, match the accuracy of humans asked to do the same task.

This result doesn't mean that algorithms are useless. Even if algorithms only equal humans, they can still offer massive advantages in terms of speed. Spotify has millions of users and employing a human operator to evaluate each of their' musical tastes would be prohibitively expensive. The data alchemists at TUI use algorithms to make sure we see the summer holiday that suits us best.

If an algorithm has the same level of performance as humans, then computers win because they can deal much more rapidly with data than a person can. So while far from perfect, models are certainly very useful.

This argument for scaling up has less validity when used to measure defendants' tendency to reoffend. The data collection required for COMPAS is complex and costly, and the number of cases processed by the algorithm is relatively small, so the argument for replacing humans with machines is weaker. There is also a big question about the invasiveness of algorithms and right to privacy of the individuals. In the Mechanical Turk trials, the workers were provided with only publically available information about the defendants, yet could make equally accurate judgements. The process of being interviewed and assessed must be demeaning for many defendants, yet it appears to provide no extra benefit in terms of predicting reoffending rates.

Of all the people I had spoken to so far, it was Julia who impressed me the most. She wasn't working for a multinational company like Google, Spotify or TUI; she wasn't employed as a professor at an academic institution like Cambridge or Stanford, and she wasn't supported by a large media outlet like ProPublica or the *Guardian*. She was an undergraduate student who wanted to challenge the structure of the world she lived in, and she had achieved amazing results in a short time. It is people like Julia who can prevent us from being outnumbered.

PART TWO

INFLUENCING US

CHAPTER EIGHT

Nate Silver vs the Rest of Us

Refresh, refresh, refresh. As voting for the 2016 US presidential election started, the political prediction website FiveThirtyEight received tens of millions of visits per hour. American voters, and citizens from around the world kept updating their browsers to find out the chances of Hillary Clinton or Donald Trump becoming the next president of the United States. The expected result, given to one decimal point of accuracy, flickered up and down – 64.7 per cent, 65.1 per cent, 71.4 per cent – and the day-by-day estimates of a Clinton win changed. The numbers had been as low as 54.6 per cent before the first presidential debate and as high as 85.3 per cent after the third, now they had finally landed on 71.8 per cent. The flickering stopped and the American people went to the polls.

I'm not sure what visitors to FiveThirtyEight thought they would find as they refreshed the decimal points of electoral prediction. I was one of them, and I'm not even sure what I thought I wanted. Some form of certainty, I suppose.

The next morning all uncertainty was gone. It was 100 per cent Trump and 0 per cent Clinton. No decimal point offered, nor required. The election had been won by the underdog and lost by the favourite.

It wasn't the first predictive failure, nor was it the most spectacular. Probably the worst polling error in recent years was made in the run-up to the UK general election of 2015. The day before the vote, the *Guardian* newspaper's model had the Conservatives and Labour neck and neck. The next morning, even the Conservative leader David Cameron seemed surprised by his parliamentary majority.

The nation's next national vote was on whether the UK should remain in the EU. The polls were close, but mostly on the wrong side of the actual outcome for Brexit. By the

2017 UK general election, even the pollsters seemed
confused as to how they should present their predictions.
Ten days before the vote, market research company
YouGov's model predicted a significant reduction in the
Conservatives share of the vote compared with the previous
election. YouGov's model contradicted other leading polls
and, after facing criticism about the manner in which it
splashed its forecast on the front page of *The Times*, it
admitted to being nervous about the outcome.[1] So when it
turned out that YouGov had got it right, and the result was
a hung parliament, it was still seen by many as a partial
failure. The perception was growing that modellers
couldn't predict elections – that the decimal points didn't
make sense.

These setbacks have occurred after a decade in which
statistical models of elections have become more and more
common. There has been a shift from newspapers reporting
opinion polls to online political sites, like FiveThirtyEight
run by Nate Silver, and The Upshot at the *New York Times,*
making probabilistic predictions of outcomes. As we have
seen, algorithms work in terms of probabilities and not in
binary outcomes. Poll predictions are no exception. Just as no
reasonable algorithm would declare that a person is certain to
commit a crime, or go to Portugal for their next summer
holiday, neither would an algorithm, even one designed for
the Huffington Post, declare that Clinton would win with
100 per cent certainty.

One challenge for the creators of these poll-based
prediction models is that us humans keep flipping their
probabilistic predictions to binary: 'yes' or 'no', 'Brexit' or
'remain' and 'Trump' or 'Clinton'. Our lazy minds like
certainty. After the 2012 US presidential election, when
Nate Silver's model predicted all US states correctly, he was
declared a genius in blogs and across social media. He was
'the man who called it right', as the *Guardian* put it. A
few years later, when Silver gave Trump a five per cent
chance of becoming the Republican party's nominee, the
same newspaper described him as making 'high-profile

misses'. By the time of the 2016 presidential election, for which FiveThirtyEight assigned a 71.8 per cent chance to a Clinton victory, social media filled with criticism of his methods. 'A rough night for the number crunchers,' wrote a columnist for the *New York Times*, implicating Silver as one of the statisticians who got it wrong.

We love to have heroes and villains, geniuses and idiots, to see things in black and white, and not in the grey reality of probabilities. The rise and fall of polls is no exception.

Let's get a few things straight from the start. Despite these failures, the models used in predicting elections are a lot better than a coin toss. Although Brexit and Trump were both assigned a less than 50 per cent chance of victory by most models, these were exceptions rather than the rule. In the majority of cases, the polls – and thus the models upon which they are based – give a probability that reflects the likelihood of the final result. A 28.2 per cent chance of Trump winning isn't small. If I threw one of my four-sided *Dungeons & Dragons* dice and got a four, we wouldn't call into question my entire role as Dungeon Master. In fact, there is evidence that better and more frequent polls are producing more accurate predictions. Figure 8.1 shows the accuracy of

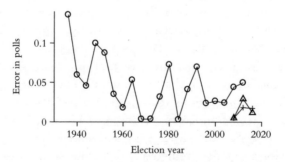

Figure 8.1 Error between poll prediction and actual outcome for the popular vote margin (relative to the sum of the major-party vote) in US presidential elections. Circles indicate Gallup polls, triangles indicate polls from RealClearPolitics and crosses polls from FiveThirtyEight.

polls for US elections over the past 80 years. The margin of error in 2016 was certainly not a sign of deteriorating accuracy.

There is a solid and well-established methodology behind the way modern election predictions are made. Poll predictors think of election outcomes as bell-shaped curves called probability distributions, as in Figure 8.2. The highest point of the bell is the most likely outcome in the election, and the width of the bell represents our uncertainty about that outcome. A very narrow bell indicates a high degree of certainty and a wide bell represents a high degree of uncertainty.

Prediction involves continually updating the shape of our bell in light of new data. This is illustrated by the curves in Figure 8.2 from top to bottom. Let's imagine that, initially, we are slightly unsure which candidate has the lead but we estimate that Clinton has a +1 point lead in the polls over Trump. We can represent our thinking by a reasonably wide bell-shape, centred on +1, as in Figure 8.2a.

A +1 point lead in the polls means that 50.5 per cent of people say they will vote for Clinton and 49.5 per cent say they'll vote for Trump. These numbers, 50.5 and 49.5, are not the probability of victory for the candidates. Instead, we can use the area under the curve to work out the probability that Trump or Clinton would win if there was an election on that day. In this case, 42 per cent of the area lies on Trump's side and 58 per cent on Clinton's. So although we think Clinton is in the lead, we still give Trump a reasonable chance of victory.

Now imagine a poll comes in that gives Trump a +1 lead nationally. One potential explanation for this poll is that Trump is genuinely ahead and our current bell has its centre at the wrong place. Another explanation is that he is still behind, but the poll happened to interview disproportionately more Trump supporters. Even the best polls contain uncertainty, both because they only contact a small proportion of the US population and because some people are still

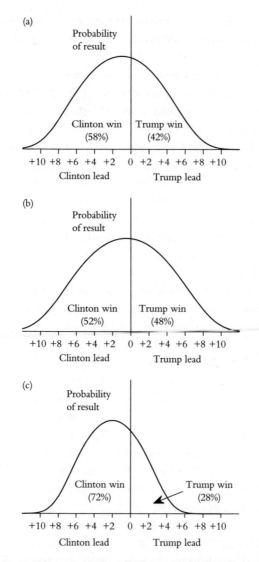

Figure 8.2a–c Three examples of bell-shaped probability distributions of outcomes. The height of the curve is proportional to the probability of different outcomes (no units). The area under the curve, left of the axis, is the probability Clinton will win overall, while area to right is probability Trump will win. Curves are for illustration purposes only and not derived from real poll data.

undecided. To reflect this uncertainty, we move the centre of our bell to the right, towards a Trump lead, and we broaden it out (as in Figure 8.2b). Something similar to this happened in the FiveThirtyEight model when an A+ rated poll by Selzer & Company on 26 September 2016, gave Trump a narrow lead. At the time, FiveThirtyEight had assigned Clinton a 58 per cent chance of winning the election. Once it had updated its model in light of the poll, it gave Clinton 52 per cent.

Over the next few weeks of the 2016 campaign, most polls gave Clinton a narrow lead and the bell moved to the left and narrowed slightly. Over time the polls slowly edged her predicted probability up towards the mid-70s. At this point, the bell-shaped curve would look something like Figure 8.2c.

The probability distribution approach requires detailed bookkeeping of all the uncertainty involved in making a prediction. Silver and his team have created a comprehensive list of political pollsters, which they rank in quality from 'A+' down to 'C-', with an additional category 'F' for polls that they believe are faking data or working unethically. Polls are weighted using this ranking and how recently they were made. FiveThirtyEight has worked out how long polls remain relevant, and uses regression models to predict how national polls reflect opinion in different states. Once all of this data is collected and combined, the FiveThirtyEight team simulates the election, accounting for a wide range of potential errors, and updates its bell-shaped curve.

The final prediction number, which so many people were clicking to update in the run-up to the election, is the area under the curve for the candidate who is in the lead.

Given the rigorous thinking behind his approach and the amount of work that has gone into weighting polls, Silver was understandably concerned with the negative coverage of his predictions after the 2016 presidential election. In a series of articles on the FiveThirtyEight website, he laid out his counter-criticism of the media. The *New York Times* was a specific focus of his ire. In the weeks leading up to the vote, the newspaper had not understood that, due to subtleties in the electoral college system, Clinton had only a very narrow lead. A *New York Times* journalist wrote that the loss of a few percentage points in the polls 'would deny Democrats a possible landslide and likely give her a decisive but not overpowering victory'.[2] In the journalist's view, victory for Clinton was more or less guaranteed. The question was simply how large it would be.

The journalist's reasoning was nonsense. All of the probability distributions in Figure 8.2 show a wide range of possible final scenarios. When a candidate has a lead of the size Clinton had – corresponding to around a 72 per cent chance of victory – many different outcomes are possible, including a Clinton landslide and a Trump victory. Emphasising one outcome or another is meaningless without discussing the underlying probabilities.

When the victory failed to materialise, the *New York Times* published an article (headlined: 'How data failed us in calling an election') that proclaimed the number crunchers had had a rough night.[3] It listed supposed problems in both their own model (the newspaper's Upshot model had given Clinton a 91 per cent chance of winning) and the approach taken by Nate Silver and FiveThirtyEight. The newspaper was blaming statisticians for its own inability to account for uncertainty. For Silver, this was just one example of how the media finds it very difficult to write sensible articles based on probabilistic reasoning.[4]

What struck me, looking at how FiveThirtyEight had evolved over the past 10 years, was that the site provides a powerful case study of the limits of mathematical models. Nate Silver had been propelled to a position of authority. He had accumulated financial resources (FiveThirtyEight is owned by ESPN) that had allowed him to build sophisticated models based on large quantities of reliable data. From reading his book, *The Signal and the Noise*, I could see that he was an intelligent and level-headed individual, who had thought deeply about how predictions work. He knew the maths and he understood the relationship between data and the real world. If anyone could create a good model of an election, it was Nate Silver.

When it came to analysing our behaviour, the algorithms we have looked at up to now were, at best, on a par with humans. Julia Dressel's Mechanical Turkers were able to predict probability of reoffending to a level similar to that of a state-of-the-art algorithm, but using a lot less data; personality models based on likes were still a long way from 'knowing us' as individuals and Spotify was struggling to come up with music recommendations as good as those of our friends.

I wondered whether the same limitations applied to the FiveThirtyEight model? Given the resources at Nate Silver's disposal, FiveThirtyEight was the undisputed heavyweight champion of algorithmic prediction. I wanted to find out whether a human contender could take it on. Could us humans match or even beat Nate's model?

During the presidential primaries, the US-based magazine *CAFE* employed a pundit, Carl Diggler, with '30 years' experience of political journalism', to make predictions for each of the US state primaries. Diggler used only his 'gut feeling and experience'. He certainly knew his stuff, calling 20 out of 22 of the Super Tuesday contests correctly.[5] As his predictions continued to pan out, he challenged Silver to a head-to-head prediction battle. Nate didn't respond, but Diggler persisted nonetheless. By the end of the primaries, Diggler had the same level of accuracy, with 89 per cent correct predictions, as FiveThirtyEight. Not only that, he had called the result of twice as many contests as Nate's site. Carl Diggler was the prediction champion of the US primaries.

Carl Diggler's predictions were real, but he isn't. He is a fictional character. Two journalists, Felix Biederman and Virgil Texas, who used their own intuition to produce the predictions, wrote his column. Their original idea was to make fun of political pundits who resemble Diggler in their pompous certainty, but when the journalists started making successful predictions, they turned on Nate. After the election, Virgil wrote an opinion piece in the *Washington Post* criticising the misleading nature of FiveThirtyEight. He accused Silver of making predictions that were not 'falsifiable' because they couldn't be tested, and criticised the way FiveThirtyEight appeared to hedge its bets using probabilities.

Virgil's criticism of Nate's methods on the basis of Diggler's success is misplaced. It is simply untrue that Nate's model is not falsifiable – it can be tested, and I'll do so over the next pages. In fact, the publication of Virgil's article in the *Washington Post* has a heavy component of what psychologists call 'selection bias' and what finance guru Nassim Taleb calls being 'fooled by randomness'. The story only made the paper because Diggler's predictions worked so well; other pundits (real or fictional) who failed to make accurate predictions were forgotten. While it is entertaining that Diggler made so many correct picks, once they are lifted out of the world of satire they have no validity whatsoever.

Predictions by so-called experts and media pundits have been extensively studied and compared with simple statistical models. Philip Tetlock, professor in psychology at the Wharton School at the University of Pennsylvania, spent much of the 1990s and 2000s studying the accuracy of expert predictions. He summarised the most striking result of this period of his research in a single sentence: 'The average expert was roughly as accurate as a dart-throwing chimpanzee.'[6] People like Carl Diggler, who work on hunches, do not outperform coin tosses in the long term. Other people, possibly like Virgil Texas, who carefully consider the data before making a prediction, tend to perform as well as simple statistical algorithms such as 'follow the recent rate of change' or stick to the status quo.

Neither Carl nor Virgil could be considered a genuine challenger to FiveThirtyEight's algorithm.

The conclusion that 'experts' usually fail was not the endpoint of Philip Tetlock's research. He followed up his study by looking at the small group of people who were able to make reasonably accurate predictions about political, economic and social events. He called these people 'superforecasters' and he found more and more of them. They came from all walks of life but had one thing in common; they gathered in and weighted information in a way that gradually improved their estimate of the probability of future events. These superforecasters would carefully adjust their forecasts in response to new information. They were humans employing probabilistic reasoning, creating bell-shaped curves in their heads.

The superforecasters were in action in the lead up to the 2016 presidential election, and as a crowd, they were just about as accurate as FiveThirtyEight. Taking the average of the prediction of all superforecasters gave an estimate of 24 per cent probability of victory for Trump.[7] This compared with 28 per cent estimated by FiveThirtyEight. Unlike Carl Diggler, who called it 100 per cent Clinton, Nate Silver and the superforecasters hedged their bets, and rightly so.

The superforecasters didn't provide individual predictions for different US states, which made it difficult to make a thorough comparison with FiveThirtyEight's predictions. So I decided, together with Alex Szorkovszky, a member of my research group, to look at how another human-based prediction method performed across all 50 states.

PredictIt is an online market run by Victoria University in Wellington, New Zealand. It allows its members to place small bets on the outcome of political events, such as 'which party will win Ohio in the 2016 presidential election?' or 'how many @realDonaldTrump tweets will mention "CNN" from noon 7/6 to noon 7/13?' Shares in events are traded directly between its users, with the price of an outcome reflecting the probability of that event. If the price for the market that Trump will tweet about CNN more than five times in that week is 40 cents, and I believe that the probability of him making six or more tweets is more than 41 per cent, then I can buy into the option. I pay 40 cents and if Trump does make more than five CNN tweets my investment becomes one dollar. If he makes five or fewer, I lose my investment.

The PredictIt market allows its users to trade in probabilities. There is no guarantee that all of its users are as smart as the superforecasters, but those that don't carefully consider the probability of events will lose money quickly. Unlike bookmakers motivated by financial profit, the website does not go to lengths to try to encourage losers to play more, or to ban winners. Their aim is to get the best forecasters betting against each other. The PredictIt algorithm pays people who make good predictions and punishes those who do badly. It is a beautifully simple way of bringing together our collective wisdom.

In his *Washington Post* article, Virgil Texas accused FiveThirtyEight of making predictions that can't be falsified. He called probabilistic predictions – such as that the chance that Clinton wins the Democratic primary is 95 per cent – 'untestable assertions'. This is true of any single electoral race; there is no way of rerunning the Brexit vote or the US presidential election in exactly the same way as they were run in 2016. But it isn't true when multiple predictions are made over many years and many different elections, as FiveThirtyEight does for US states. If FiveThirtyEight declares that it is 95 per cent certain on 10 different events and fewer than half of these events occur, then we might well question its methodology. If nine or 10 of these events occur then we are more likely to accept the method is sound.

A good way of visualising the quality of predictions is to group them together on the basis of how 'brave' they are and then compare them to the proportion of time the predicted outcome occurs. I do this in Figure 8.3 for both FiveThirtyEight and prediction markets Intrade and PredictIt.[8] The brave predictions are the ones that lie at the extreme left and right of the figures. These correspond to a prediction that the Democrat candidate will win or lose a state with a certainty greater than 95 per cent. All of these brave predictions, for both FiveThirtyEight and the prediction markets, proved to be correct: the strong favourite won the state.

More cowardly predictions, where certainty of the prediction lies between 5 per cent and 95 per cent, are shown

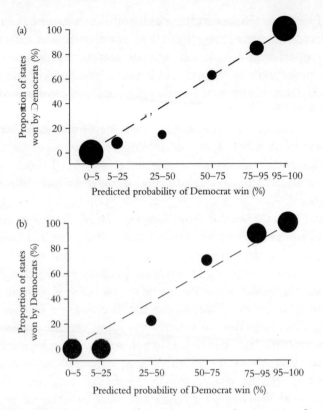

Figure 8.3 Comparison of predictions and outcome for (a) FiveThirtyEight and (b) prediction markets Intrade/PredictIt for all US states for presidential elections in 2008, 2012 and 2016. Radius of circle is proportional to number of predictions in each case.

in the interior of Figures 8.3a and 8.3b. The quality of these predictions can be measured by their distance to the dotted line. Circles above the line indicate that the predictions typically underestimated the Democrats' chances, while circles below the line indicate that the Democrats' chances were overestimated. There was a very slight tendency in 2016, by both prediction markets and by FiveThirtyEight, to overestimate the Democrats' chances in states where the Republicans were favourites. This tendency is not statistically significant and can be reasonably attributed to chance.

The quality of predictions can also be measured in terms of the number of 'cowardly' and 'brave' predictions. If I had declared in advance of each presidential election since the Second World War that there was a 50 per cent chance of a Democrat winning, my predictions would be correct half the time, simply because about half of the presidents have been Democrats. But I don't think anyone would declare me a genius for calling every election as a 50/50 coin toss.

The size of the circles in Figure 8.3 is proportional to the number of each type of prediction. So, the large circles at the bottom and top of the FiveThirtyEight plot show that they make many more 'brave' predictions than the 'cowardly' predictions that lie in the middle. Prediction markets are slightly less 'brave', especially when it comes to strong favourites. This is probably because of what is known in gambling as a 'longshot bias' – there is always someone who fancies taking a gamble by buying a five-cent prediction of a very unlikely event.[9] These people will nearly always (with a probability bigger than 1 in 20) lose their money.

By calculating a measure known as the Brier score,[10] my colleague Alex and I found that over the three election years there was very little difference in the performance of state-by-state predictions of the model and the crowd. FiveThirtyEight was slightly better than the prediction markets in 2012 when it made a very brave prediction that Obama was a clear favourite. The wise crowd of humans on PredictIt and the model at FiveThirtyEight were equally good in their assessment of the 2016 presidential election. Both gave similar opinions on most states and both got Trump less wrong than the impression given by much of the media.

Carl Diggler and his gut-feeling predictions came in third place, but not as far behind as I thought he might. His Brier score for the 2016 state-by-state predictions was 0.084, compared to 0.070 for FiveThirtyEight and 0.075 for PredictIt (lower Brier scores are better). I have to give Virgil Texas some credit, he wasn't as bad an 'expert' as some of the chimpanzees with darts that Philip Tetlock had studied.

There is a problem, though, in comparing PredictIt and
FiveThirtyEight. While they both make reasonable
predictions, they are not independent.

If we follow how the two predictions changed over time, we
see that they lie very close together (Figure 8.4). This can be
partly explained by PredictIt users exploiting FiveThirtyEight.
Indeed, on the superforecasters' forum discussions, the single
most frequent source for information was Nate Silver's website.
However, the whole point of a prediction market is that it
brings together different pieces of information, weighing them
in proportion to their quality. So while it would be considered
as high quality, it is unlikely that FiveThirtyEight was the sole
source of the PredictIt market. And there is certainly no
evidence of PredictIt predictions following the ups and downs
of FiveThirtyEight with a delay.

FiveThirtyEight doesn't explicitly use betting market data
in its model. However, Silver, a former professional gambler,
understands very well that prediction markets and bookmakers'
odds give a better reflection of the probability an event will
happen than the polls themselves. He could see that the
markets were not very certain about a Clinton victory. Other
models, at the *New York Times* and the Huffington Post, which
were based purely on polls, were predicting a 91 per cent and
99 per cent, respectively, win probability for the Democrat

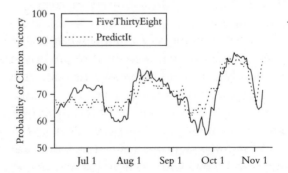

*Figure 8.4 How the predicted probability of a Clinton victory changed
in the months leading up to the 2016 US presidential election for
FiveThirtyEight (solid line) and PredictIt (dotted line).*

candidate. The FiveThirtyEight team applied an adjustment to the polls in order to reflect uncertainty about the outcome, bringing it nearer to the market odds.

While this adjustment turned out to be justified in terms of getting the election less wrong than his competitors, the fine-tuning raises an issue about the basis of his approach. Former FiveThirtyEight writer Mona Chalabi told me that Nate's team would use phrases such as 'we just have to be extra cautious', to express a shared understanding within their newsroom that the model shouldn't give too strong predictions for Clinton. They were aware that they would be judged after the election in the same black-and-white terms that humans always judge predictions: they would either be winners or losers.

Mona, who is now a data editor at the *Guardian US*, told me: 'The ultimate flaw in FiveThirtyEight, and all election forecasting, is the belief that there is a method to correct for all the limitations in the polls. There isn't.' Academic research has shown that polls are typically less accurate than prediction markets.[11] As a result, FiveThirtyEight has to find a way of improving its predictions. There is no rigorous statistical methodology for making these improvements; they depend much more on the skill of the individual modeller in understanding what factors are likely to be important in the election. It is data alchemy: combining the statistics from the polls, with an intuition for what is going on in the campaign.

Mona made this point very strongly when I talked to her: 'The polls are the essential ingredient to prediction and the polls are wrong. So if you take away the polls, how exactly are they going to predict the election?'

FiveThirtyEight is an almost entirely white newsroom. They are, for the most part, American, Democrats and male. They have followed the same courses in statistics, and share the same world view. This background and training means they have very little insight into the mind of the voter. They don't talk directly to people to get a sense of the feelings and emotions involved, an approach that would be considered subjective. Instead, Mona described to me a culture where colleagues judged each other on how advanced their mathematical techniques were. They believed there was a direct trade-off between the quality of statistical results and the ease with which they can be communicated.

If FiveThirtyEight offered a purely statistical model of the polls then the socio-economic background of their statisticians wouldn't be relevant. But they don't offer a purely statistical model. Such a model would have come out strongly for Clinton. Instead, they use a combination of their skills as forecasters and the underlying numbers. Work environments consisting of people with the same background and ideas are typically less likely to perform as well on difficult tasks, such as academic research and running a successful business.[12] It is difficult for a bunch of people who all have the same background to identify all of the complex factors involved in predicting the future.

I can't see how Silver and his team can beat the efficient market of PredictIt in the long term. I haven't tried to predict election results myself, but I have done a bit of betting on football (purely for scientific research reasons, of course). There is an urban legend of the mathematical genius, maybe the Nate Silver of gambling, who has worked out the formula for beating the bookies. If only, the legend goes, you can find the tips that this person can provide, the source of the magic equation, you could become rich beyond your wildest dreams.

This legend is pure mythology. There is no equation that predicts the outcome of sports events. The only way to make a profit betting on football is to incorporate the odds provided by the bookmakers into your mathematical model. This is what I did in my last book *Soccermatics*, when I created a betting model. I used statistical patterns in the odds to identify a small but significant bias in how they were set and exploited this bias to make money. There is mathematics involved in the process of modelling football, but a gambler who thinks they can beat the market without incorporating the wisdom already held within the betting crowd, is going to lose eventually.[13]

This same 'you can't beat the bookies without playing their game' logic applies to Nate's work. He admits that his sports models don't beat the bookmakers' odds. Punters incorporate both the FiveThirtyEight predictions and other relevant information into the market price. They always have the edge over one individual, however clever they happen to be.

Mona learnt a lot from her experience at FiveThirtyEight, but not in the way she had imagined when she started. She went into the job hoping to develop her skills in data journalism but came out understanding that the accuracy offered by FiveThirtyEight was an illusion. It was Mona who alerted me to the decimal points in FiveThirtyEight's predictions. I hadn't thought so much of them before. I saw probabilities like 71.8 per cent as a number, but I had forgotten that they also imply precision. In science lessons in school, we learn to use significant figures to reflect the degree of accuracy with which we can measure a quantity. For example, if we know that 10 bags of sand weigh between 12.6kg and 13.3kg, we can say that a bag of sand weighs 13kg, to two significant figures. All opinion polls have an error of at least three percentage points, and usually more. This error means that, at the very best, the election prediction probabilities should be stated with only one significant figure, e.g. 70 per cent. Any more decimal points than this are misleading.

For all their advanced mathematics, FiveThirtyEight are making schoolboy mistakes in their rounding. I can thoroughly recommend the website BBC Bitesize if you, or they, would like to learn more.

This understanding is something that Mona has taken into account in her more recent work as a data editor at the *Guardian US*. She uses a small number of categories when presenting data and illustrates it in a way that emphasises the uncertainty in the measurements. She concentrates on ordering data, rather than just relying on numbers, and she never uses decimal places. One of her drawings that struck me most was of the area of a parking space and a solitary confinement prison cell drawn side by side. I don't remember how big each of them was, but the cell was much smaller than the parking space.

Before I spoke to Mona, I have to admit that I saw assessing FiveThirtyEight as a game. I thought it was fun to compare models to markets, to see which won. But I was falling into the same trap as Nate Silver. I was forgetting that the outcome of the US presidential election was crucial to the lives of so many people. Mona told me that the real danger is that, 'FiveThirtyEight potentially influences voter behaviour. The millions of people visiting the site on election day aren't reading through Nate's model to find out how it works. These are people who are looking at those two numbers for Clinton and Trump, and they are concluding that Clinton is going to win.'

We are outnumbered by statistical experts like Nate Silver because we believe that they have a better answer than we do. They don't. They might be better than chimpanzees with darts and they might narrowly beat a (pretend) pundit like Carl Diggler, but they don't beat our collective wisdom. If you are interested in finding out more about the intricacies of creating models, then I thoroughly recommend the FiveThirtyEight pages (rounding error excepted). If you are just checking the headline number for the next election, you are wasting your time. Use the bookmakers' odds instead.

Inspired by Mona's emphasis on giving rough categories for differing outcomes, I propose an obligatory warning to all model predictions, phrased in a way everyone can understand. We could have three categories of performance:

Random: these predictions do not outperform chimpanzees with darts.
Low: these predictions do not outperform an underpaid
Mechanical Turk worker.
Medium: these predictions do not outperform the
bookmakers' odds.

This covers the performance of most mathematical models used for predicting human-related events in sport, politics, celebrity gossip and finance on the timescale of days, weeks and months. If you do know of a reliable exception to this rule, please tell me ... I'll be straight down the bookies to back the model.

CHAPTER NINE

We 'Also Liked' the Internet

The FiveThirtyEight algorithm and the PredictIt algorithm were different from those I had looked up to that point. They didn't just classify us. They interacted with us. The FiveThirtyEight model influenced us. It's difficult to know whether, as Mona Chalabi suspects, it influenced whether or not people voted, but it certainly influenced how Americans felt about the upcoming election.

We interact with algorithms from the second we open our computers or switch on our phones. Google is using the choices of other people, and the number of links between pages, to decide what search results to show us. Facebook uses our friends' recommendations to decide the news we see. Reddit allows us to 'up vote' and 'down vote' celebrity gossip. LinkedIn suggests the people we should meet in the professional world. Netflix and Spotify delve into the film and music preferences of its users to make suggestions for us. These algorithms all build on the idea that we can learn by following the recommendations and decisions made by others.

Is that really where we are now? Do the algorithms we interact with online really provide us with the best quality information?

Jeff Bezos, founder of the retailer Amazon, was the first person to recognise that we only want to see a small number of relevant choices when browsing. His company introduced terms like 'related to items you've viewed' and 'customers who bought this item also bought' to help us find the products we want. Amazon gives us a small number of choices from many millions of different options. 'You've read *Freakonomics*? Try *The Undercover Economist* or *Thinking, Fast and Slow*'. 'You've viewed Jonathan Franzen's latest novel? Most customers went on to buy *A Little Life* by Hanya Yanagihara'. 'Frequently bought together: Kate Atkinson, Sebastian Faulks

and William Boyd'. 'You searched for *Soccermatics*? Also try
The Numbers Game and *The Mixer*'. These suggestions give an
illusion of choice, but the books have been clustered together
by Amazon's algorithm.

The reason this algorithm is so effective is that it
understands us. When I look at the books suggested for my
favourite authors, the recommendations are spot on. Either I
already own the book, or it is one I would like to get my
hands on. During the two hours I just spent on Amazon's
website 'researching' their algorithms, I ended up putting
seven items in my basket. The algorithm understood not just
me, but also my wife and my relatives. I just did all my
Christmas shopping in one sitting. It even understands my
teenage daughter better than I do: when I looked up Dodie
Clark's book *Obsessions, Confessions and Life Lessons* it suggested
that Elise might also like *Turtles All the Way Down* by John
Green. I am sure she will.

When I read fiction, I hear another person's words in my
own voice. It is a very personal experience, a special
connection between me and the writer. Sometimes when I
am deep in a good novel, I believe that no other person will
ever talk to me in the way this writer has talked to me.

A few hours on Amazon dispels this illusion entirely. The
algorithm uses the fact that other people who like the books
I do, have made choices similar to the ones I might make. To
categorise its 10 million products, Amazon creates links
between the things that customers buy together. These links
are then the basis for the suggestions they make to us. This is
simple but effective. In Amazon's customer base, there are
lots of people like me, like my children and like my friends.
It turns out an algorithm, developed by a small group of
researchers in California, can simply pick out a new special
voice for me to listen to and dispatch it to me, free of charge,
with next-day delivery.

I don't have access to the exact details of how Amazon's
current algorithm is constructed: the secrets are held within
Amazon's daughter company, product search specialists A9.
The algorithm changes over time and varies for different

products, so there isn't one single 'Amazon' algorithm anymore. There are, however, basic principles behind the Amazon algorithm and how we interact with it, and these can be captured in a mathematical model.

In honour of Amazon, I call this model 'also liked' and I'll now go through the steps involved.[1] In my model, there are a fixed number of authors. I'll take 25 popular science and maths authors because it's fun to use recognisable names, but the names don't affect the outcome of the model. At the start, I assume that no purchases have been made, so the first customer buys two books at random and each author is equally likely to be chosen.

From then on the modelled customers arrive one at a time. Each customer is more likely to buy books by pairs of authors who have previously been bought together. Initially, this effect is quite weak. The general rule is that the probability an author's book is bought is proportional to the number of previous purchases plus one.[2] The 'plus one' ensures that every book has a chance of being purchased. Imagine, for example, the first customer bought books by Brian Cox and Alex Bellos. The probability the new customer buys Cox or Bellos is 2 in 27 respectively, while the probability they buy any other author is 1 in 27.

In Figure 9.1a, I show the first 20 purchases in a simulation of the 'also liked' model. A line between two authors indicates that a customer has bought books by both of them. The first customer's random choice of Cox has led to four more joint purchases. Ian Stewart has been bought four times. Richard Dawkins and Philip Ball have sold three books each. At this stage, it is not clear which author is most popular.

After 500 sales the picture is very different. Figure 9.1b shows that Steven Pinker is by far the most popular author, with strong links to Daniel Kahneman, Susan Greenfield and Philip Ball, who are also selling well. Richard Dawkins and Brian Cox have fallen behind, and several other good authors have failed to take off.

In the simulation, some authors become very popular, as more and more connections are made to them, and other

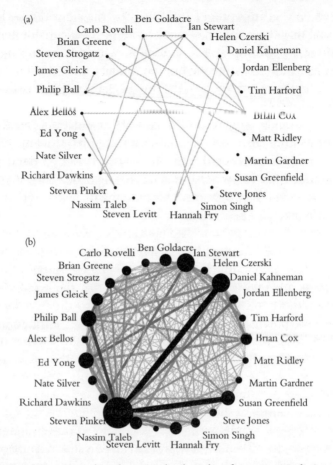

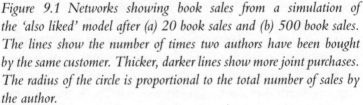

Figure 9.1 Networks showing book sales from a simulation of the 'also liked' model after (a) 20 book sales and (b) 500 book sales. The lines show the number of times two authors have been bought by the same customer. Thicker, darker lines show more joint purchases. The radius of the circle is proportional to the total number of sales by the author.

authors sink into obscurity. The combined sales of the top five authors are approximately equal to that of all the other 20 authors put together.

For the authors, it is here that the potential danger in the also liked algorithm lies. The customers in my model do not take into account how good the book is: they buy books based on the links provided to them by the algorithm. This means that the number of books sold by two equally good authors can end up at two extremes, with one becoming a bestseller and the other selling far fewer copies. Despite all books being of exactly the same quality, some are bestsellers, others are flops.

Kristina Lerman, researcher at the Information Sciences Centre of the University of Southern California, told me that our brains love 'also liking'. She uses a rule of thumb to model how we behave online. She told me: 'Basically, if you assume people are lazy then you can predict most of their behaviour.'

Kristina drew her conclusions after studying a very diverse set of websites, from social networks like Facebook and Twitter to programming sites like Stack Exchange, online shopping at Yahoo, the academic network Google Scholar, and online news sites. When we are offered a list of news articles, we are much more likely to read those at the top of the list.[3] In a study of the programming question–answer site Stack Exchange, Kristina found that the users tend to accept an answer based on how far up it is on the page and how much space (not number of words) it takes up, rather than the apparent quality of the answer.[4] Kristina told me: 'The more options people are shown on sites like these, the fewer options they look at.' When we are shown too much information, our brains decide that the best thing to do is just ignore it.

Kristina told me that 'also liking' generates 'alternative worlds', where popularity online is decided by lots of people who are not thinking particularly hard about the choices they are making and reinforcing other peoples' poor decisions. To better understand these alternative worlds I ran a new simulation of my 'also liked' model, with the same 25 authors. Since the algorithm is probabilistic, no two outcomes of the simulation are exactly the same. In the new outcome, shown in Figure 9.2, it is Martin Gardner who builds on early attention and becomes an international bestseller. Each run of the simulation generates its own unique bestseller list. In each simulated world, early sales are reinforced and a new

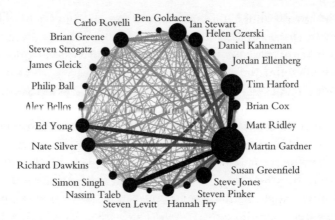

Figure 9.2 Networks showing book sales from a new run of the simulation of the 'also liked' model after 500 book sales. The lines show the number of times two authors have been bought by the same customer. Thicker, darker lines show more joint purchases. The radius of the circle is proportional to the total number of sales by the author.

popular-science writing success is born. Also liked creates success at random.

We can't rerun the real book market in order to test how much success is explained by an initial lucky break and how much is explained by quality. Once certain authors are established then it is impossible to remove them from history. I don't think successful popular-science authors, like Ben Goldacre and Carlo Rovelli, would be too thrilled if Amazon had suggested resetting their book sales directly after they released their books in order to do an experiment on the alternative worlds theory of book sales. Nor would Beyoncé, Lady Gaga and Adele want iTunes and Spotify to reset their top lists in order to test an alternative worlds theory of popular music.

While we can't rerun our own musical world, it is possible to create smaller artificial worlds. Sociologist Matthew Salganik and mathematician Duncan Watts performed an experiment in which they created 16 separate online 'music worlds' where users could listen to and download songs from unknown bands.[5] The songs were the same in each world, but each world had its own chart. The chart showed the number of downloads by other members of a user's world, but not downloads from other worlds. Matthew and Duncan found that songs at the top of the chart were about 10 times more popular than those found mid-chart. And the charts looked different in different worlds, with chart-toppers in one world languishing in the middle of the chart in others.

The researchers also measured how good the songs actually were, by observing how listeners who received no chart information rated them. Songs that these independent listeners liked did better in the charts than those songs that the listeners didn't like. But it was still difficult to predict which song would top the charts. Really bad songs never made it, but good and OK songs were equally likely to become hits. It seems that everyone knows a bad song when they hear it, but an artist doesn't need to be exceptional to succeed.

Being 'liked' is worth a lot to individuals, businesses and media outlets. Sinan Aral, at MIT Sloan School of Management, has started to quantify the impact of 'liking'. He and his colleagues worked together with a popular news-aggregator site to find out what happened when they up- and down-manipulated posts.[6] Adding a single extra 'upvote' to a post directly after it was published elicited further 'upvoting' by other users. Once all voting had finished, the effect of the initial upvote could still be seen in the total votes, providing half an extra vote on average to the post. The effect is small but demonstrates that vote manipulation works.

Sinan's study revealed a key difference between the PredictIt algorithm and the 'also like' algorithm. In PredictIt there is always an incentive for users to oppose the prevailing wisdom, because of the possibility of making a financial profit. Users of news aggregators don't oppose positive judgements; instead, they become more inclined to 'upvote' themselves. In contrast, when posts were manipulated with a 'downvote', other users were quick to counter with an 'upvote'. In this case, the manipulated 'downvote' had no overall effect on the final ranking. We regulate negative judgements but uncritically approve of positive judgements. Our brains might be lazy, but at least they tend to prefer positivity over negativity.

Sinan had agreed not to reveal exactly which news-aggregation site he worked with, but the leading site of this kind just now is Reddit. Erik Martin, the general manager of Reddit at the time the study was published, told *Popular Mechanics* magazine that Reddit had caught several publishers trying to systematically manipulate the site.[7] Reddit has algorithmic bots that patrol its pages looking for fake accounts posting in unhuman-like ways. Erik said: 'We have counter-measures and people who actively look for that stuff, a community that has no tolerance for that manipulation.'

Reddit works well because humans closely monitor its most-read threads, but it is impossible to have humans monitor and control every part of the Internet. And that leaves opportunities for exploiting the 'also like' Internet.

I tracked down an old friend who I knew could tell me more about these possibilities. My friend wanted to be identified only as 'CCTV Simon', his online persona and not his real name. After completing a master's degree in Informatics, Simon became a stay-at-home dad, resisting the temptation to follow his fellow graduates to jobs at Google and other tech companies. Simon had a bit of time on his hands between nappy changes and was thinking about ways he could use it to make some cash from home. It was then that he discovered BlackHatWorld.

If you are interested in buying a new camera, you'll probably read a few online reviews before going on to make your purchase. When you have found out everything you need to know, you'll then go to Amazon or another leading retailer to make the purchase. Often your route in to Amazon will involve clicking on an advert or following a link offered by the sites containing the reviews and information. These intermediate sites, known as gateways, can apply for Amazon affiliate status. For every purchase that originates from the gateway, Amazon pays out a small commission to these affiliates. For large, established websites this is a good source of advertising revenue. BlackHatWorld is a forum for affiliates who want to make money, but don't want the hassle of creating a website with genuinely useful or interesting content.

The term 'black hat' was originally applied to hackers who broke into computer systems and manipulated them for personal gain. The affiliate black hats don't hack into Google, but they do everything they can to exploit Google's search algorithm to make money. CCTV Simon realised that if he could create an affiliate website that topped a few of Google's search results, then a lot of traffic would pass through his site on the way to Amazon. As Kristina Lerman has shown, it is the top result that our lazy brains are interested in. With the help of posts on the BlackHatWorld forum, Simon developed a strategy. He decided to focus on CCTV cameras, because they are a growth area in the UK and are sufficiently expensive that the commission could earn him money. He studied Google AdWords and came up with a few key search phrases to put into his page. By titling one page: 'Top 10 mistakes you can make when buying a CCTV camera', he found a gap in the market: no other black hat affiliates had exploited that particular search term.

The next stage is to fool the Google algorithm into believing that people are genuinely interested in an affiliate site. Simon calls this 'creating the link juice'. In Google's original PageRank algorithm, a site was ranked in terms of the flow of clicks through it, which depended in turn on the numbers of in and out hyperlinks it had. The Google algorithm builds on the same principle as 'also liked' – the more popular a page is, the more likely it is to be shown to others when they search for a subject.[8] As a page climbs the ranks, the traffic to the site increases and its position is further enhanced.

Black hat affiliates manipulate Google's results by creating multiple links to the pages they want to promote. When Google's algorithm looks at connections into a site, it thinks that the linked-to site is central to the network and moves it higher up its list of search results. Once the link juice is flowing, and a site is making it to the top of search lists, still more juice is generated as real users start clicking through. This is when black hats start to make money, not from Google, but from the clicks that feed into Amazon and the other affiliate sites that pay commission.

Over time, as Google has developed ways of spotting bogus links, the black hats have gone to greater lengths to fool the algorithm. Currently, the popular approach is to create 'private blog networks', where one person sets up 10 different sites about, for example, widescreen TVs. The operator hires ghostwriters to fill the sites full of more or less meaningless text on the subject. These 10 sites then link to an affiliate site, making it look as if the affiliate is the top authority on modern televisions. Lone black hat operators are creating entire online communities – complete with Facebook likes and Twitter shares – just to fool Google's algorithm.

A successful private blog network or black hat Amazon affiliate site does have to have some genuine written content. Google uses a plagiarism algorithm to stop sites simply copying other sites and it uses automated language analysis to make sure the articles on the site follow basic rules of grammar. Simon told me that he had initially bought the CCTV cameras and wrote genuine reviews. But then he realised that 'Google doesn't give a shit if I bought the camera or not. All its algorithm is doing is checking for keywords, looking for original content, seeing if I have a few pictures and measuring the juice.' Simon knew from his university studies how the algorithm worked but was amazed at just how crude Google's approach to traffic was. Soon his site was attracting hundreds of thousands of views.

Affiliate sites vary immensely. Simon contrasted his own site with the 'white hat' affiliates, who he described as, 'very genuine, very personal, home moms in the US who have pictures of themselves and really care about the products they present'. There is also a grey area, occupied by sites like HotUKDeals. This site encourages its members to share tips about leading retailers, giving the impression of a community of bargain hunters. While the site does have a large genuine user base, I found that HotUKDeal posters were also being recruited on BlackHatWorld in order to post for certain affiliate sites. Simon believed that most users were oblivious to the site's real purpose: every single link out of HotDealsUK is one of the site's affiliates, so every tip generated cash for the site's owners.

At the peak of Simon's operations, he was making £1,000 per month from a website containing made-up reviews and meaningless tips. Browsing the site, I was impressed by the writing style he had mastered: a kind of *Top Gear* review style for CCTV that didn't really say anything. The site poses questions and considerations back to the reader, such as 'want a simple budget indoor IP cam?', or 'make sure you do your homework if you want to keep an eye on baby'. It 'explains' some of the jargon and it provides long reviews that skirt the issue of whether or not anyone at the site has actually used the camera.

Although his CCTV site still brings in a couple of hundred pounds a month, Simon is no longer maintaining and updating it. He did consider investing and building up a larger number of affiliate sites, but thought, 'could I then put my kids to bed at night and tell them that Daddy had done a good honest day's work?' His answer was no, and when he returned to the workforce, he took a proper job instead.

Searching on Google, I find that all of the top five suggestions for 'home CCTV camera' have embedded links to Amazon. None of the 'reviews' give any indication that the writer has actually used the products. *Which?* magazine is in seventh place on the Google ranking and claims to have proper reviews, but they were hidden behind a pay wall. In 20th place, the *Independent* had some good reviews, with relatively unobtrusive links to the manufacturers.

When commercial interests are involved, the collective reliability of Amazon and Google can drop dramatically. The positive feedback created by 'also liking', and the importance Google put on traffic means that a genuine white hat affiliate – that really did systematically review all widescreen TVs or

CCTV cameras – would quickly disappear among all the black hat sites that are just pumping their juice. Unlike the PredictIt algorithm, where the financial incentive for the punters is to make better predictions, the financial incentives around buying and selling products online lie in increasing uncertainty for the consumer.

With all of the distortion online, I couldn't help thinking a little about my own personal success. How is *Outnumbered* going to do when it is released? I am not going to create an army of bots to click on Amazon links or generate a community of people recommending the book on the website Goodreads. Is there any chance of my book becoming a bestseller?

I showed my also liked simulation results to sociologist Marc Keuschnigg. He has studied book sales in detail, trying to find the secret to what makes a best-selling book. He agreed with me that a large part of the explanation of success or otherwise of a book lies in how much it was also liked on Amazon. But not all books and authors are equal.

'There is a big difference between newcomers and established authors. It is newcomers who are most subject to the peer effect,' he told me. 'When people don't have information about which book to buy, they look to see what their peers have bought.'

Marc studied sales of German fiction in bookshops from 2001–06, just before Amazon came to dominate the market, and found that if a newcomer got on to the bestseller list then its sales were boosted by a further 73 per cent in the following week. The visibility of getting into the top 20 further increased sales.

Visibility in top lists was only part of the story. Another factor that helped newcomers was bad reviews in the press. Yes, that's right, not good reviews, but bad ones. A negative review in a newspaper or magazine typically produced a 23 per cent boost in sales for books written by first-time novelists. Positive reviews, on the other hand, had absolutely no effect. Marc told me: 'There is a high risk that the bestseller lists contain mostly average and bad-quality books.'

To back up this rather strong claim, Marc showed me an analysis he had performed relating sales to online reviews. The more copies a book sold, the fewer stars it received on Amazon. This may well be a form of revenge by disappointed readers.[9] They see a book climb the charts or appear in an also liked list and are persuaded to buy it. When the book proves to be boring, they take out their frustration by awarding it a low score on Amazon. It seems that readers never learn: we ignore the reviews and go with the crowd. It is only after the crowd gets it wrong and the negative feedback kicks in that the reviews start to reflect the quality of the book.

From my point of view, the role of 'also liking' in distorting quality, makes success bittersweet. I was pleased when my last book, *Soccermatics*, sold reasonably well, but I also got a one-star review on Amazon that made me think about Marc's study. This reader had bought the book because he had read on a betting forum that it was a manual for taking money from the bookmakers. Although this was never my intention, the reviewer was sadly disappointed. He (gender assumed) wrote: 'Ninety-nine per cent of reviews on here are fake, I have read the book, very very poor, you will waste your money on this, you will not improve any chance of winning bets reading this book and only line this guy's pocket.'

When we know we are living in one of many alternative worlds, success in our real world can feel very hollow.

CHAPTER TEN

The Popularity Contest

D uring the summer of 2017, my son played a truly terrible song for me: a rap called 'It's Everyday Bro'. Accompanied by California hip-hop beats, it starts with a line that made me feel physically ill: 'It's everyday bro with that Disney Channel flow'. It then goes on to brag about the number of followers the artist, Jake Paul, has on YouTube, before he hands over to his 'Team 10' posse to break it down with lines like: 'My name is Nick Crompton ... Yes, I can rap, but no I'm not from Compton'.

I know, and my son has confirmed, that the song is meant to be somewhat ironic. But Jake Paul reveals the sheer extent to which popularity has become a warped form of 'also liking'. Jake rose to fame through the video site Vine, and then after an acting role on the Disney Channel he launched his YouTube channel. Here, we can watch him driving his 'Lambo' past his old school, we are taken on a tour of his luxury mansion in Beverly Hills and see him shouting out of Italian hotel room windows. In all his films, songs and comments on social media, he encourages his followers to 'like' and share everything he does.

An original aspect of Jake Paul's popularity explosion is that he has found a way of exploiting YouTube 'dislikes' to the same extent as 'likes'. A unique selling factor of his 'It's Everyday Bro' video, is that it received two million dislikes, more than any other video ever uploaded. That's 'never done before', as Jake Paul would put it. Kids are watching the video and the 10-second advert that precedes it, just to dislike it. Paul is popular because he so mediocre, so self-absorbed and so unashamed in his craving for popularity.

During the second half of 2017, many YouTubers – whose main online activity up to that point was to film themselves

playing computer games or performing pranks on their
friends – started hiring music producers and rap artists to help
them make 'diss tracks' about other YouTubers. A leading
'star' in this movement is RiceGum. He specialises in making
fun of the videos other people upload, often in the form of
raps, in the hope that they 'diss' him back, sending more
traffic to his channel. Much of the content of the dissing and
counter-dissing refers explicitly to how 'also liked' they are,
and how much money they have. When RiceGum dissed
Jake Paul's mansion tour video, Jake responded with a long
rant about how RiceGum's Lamborghini 'was hired' and that
he 'only made $60,000 a day'. Amazingly, the more references
these YouTubers make to being rich and liked, the more kids
follow them, subscribe to their channel, watch endless adverts
and buy their 'merch' (merchandise).

There is nothing particularly unique or special about Jake
Paul or RiceGum. They are young men who have been
lucky enough to propel themselves to the top of every
YouTube must-watch list. As they get more 'likes' and more
'dislikes', more advertising revenue and more sales on iTunes,
they gain more attention and even more success. They are a
product of an algorithm that rewards attention-seeking and
shock value.

Charts, prizes and the advantages that come with
accumulating success and contacts have always been a part of
our lives. Sociologists have long known about the 'rich get
richer' effect. Like CCTV Simon, Jake Paul and RiceGum
understand the importance of click juice and being 'also
liked', they explicitly encourage their army of fans to get
involved. The 'up next' videos of YouTube amplify the effect.
Every watching, listening or purchasing decision their fans
make is accompanied by a small snippet of information about
the choices made by others. Through a series of clicks and
suggestions, teenage and pre-teenage girls and boys create a
world of superstars whose success partly reflects the quality of
their work, but also reflects a desperate need to make sure
they aren't missing out. For Jake Paul, the journey from
relative obscurity to world domination took just six months.

It can happen to anyone with a modicum of talent, but it doesn't happen to many of us.

In academia, the equivalent of making it to the top of YouTube subscriptions is to make it to the top of the citation lists on the Google Scholar service. Like YouTube, Scholar provides just one simple feature: a list of links to articles related to the search term that was entered. The order in which the articles are presented is determined by the number of times they had been cited (or referred to) by other articles.

Citations are essential to academia because they create our discourse. References inside an article and a list of citations at the end show how the article contributes to a joint understanding of problems. The number of citations a paper attracts is a very good way of assessing an article's importance in a field. The more citations an article has, the better it represents how scientists are thinking.

Ordering articles by citations makes sense, but Google Scholar's approach has an unexpected side effect. When the website first started in 2004, the journal *Nature* interviewed neuroscientist Thomas Mrsic-Flogel. He told them: 'I follow the citation trail and get to papers I hadn't expected.'[1] Instead of visiting the library, or even the web pages of scientific journals, he used citation links between papers to find new ideas. Just as my son browses backwards and forwards between YouTubers, Thomas was browsing backwards and forwards between scientists.

I'm not judging. At that time I was doing exactly the same thing as Thomas. I still do it today. I browse backwards and forwards through articles, clicking near the top of the list, trying to work out what is going on in my research field and who is publishing the best stuff. And so do all of my colleagues, too. Very soon after Scholar came out we were all addicted.

The Google engineer Anurag Acharya, who co-created Scholar and continues to run the site, has said his original aim was to 'make the world's researchers 10 per cent more efficient'.[2] This was a highly ambitious goal, but it is one that he has already surpassed. While writing a book like this one, I make about 20–50 Google Scholar searches per day, saving me an incredible amount of time and effort. This book and my research would be impossible without Scholar.

What Anurag didn't know at the time was that he had created an also liked algorithm for academia. If a paper is cited more times, it is higher up the list of retrieved articles and more likely to be seen by other scientists. This means that popular articles are read and cited more often, leading to feedback; certain articles move up the citation lists, and others fall down. Just as with books, music and YouTube videos, the rise and fall of a scientific article's citations may have less to do with genuine quality than it does with small differences in their initial popularity.

In situations where there are a large number of items that can be 'also liked' – such as the hundreds of thousands of scientific articles that have been written – then popularity can often be captured by a mathematical relationship known as a 'power law'. To understand power laws, think about a plot of the proportion of papers that are cited more than a certain number of times. We are most used to graphs with points that increase on a linear scale, i.e. in equally spaced steps, like 1, 2, 3, 4 etc. or 10 per cent, 20 per cent, 30 per cent etc. Power laws are revealed when we plot data on a double logarithmic scale, in which we increase the steps in successive powers of a number. For example, the positive (or strictly speaking, the non-negative) powers of 10 are 1, 10, 100, 1,000, 10,000, etc. Similarly, the negative powers of 10 are successively smaller – 0.1, 0.01, 0.001, etc. On a double logarithmic scale, the x-axis is how many times a paper has been cited and the y-axis is the proportion of papers that have been cited that number of times or more.

A double logarithmic plot for scientific articles in 2008 is shown in Figure 10.1. There is a straight-line relationship between the proportion of articles and number of citations (for articles cited more than about 10 times). It is this straight line that is known as a power law.[*]

---

[*] The proportion of articles that are cited p times or more is proportional to the number of times n an article is cited to the power of a constant number $-a$ (i.e. $p=kn^{-a}$ where k is also a constant[4])

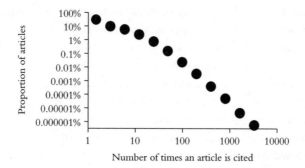

Figure 10.1 *The number of times an article is cited plotted against the proportion of articles cited more than that number of times for scientific articles in 2008. Data collected by Young-Ho Eom and Santo Fortunato.*[3]

Power laws are a sign of vast inequality. In 2008, 73 per cent of scientific articles had been cited once or less. A very depressing thought for anyone who has spent those many months required to write an article. At the other end of the spectrum, one in 100,000 articles were cited 2,000 times or more. There are lots of unsuccessful articles, that no one ever reads, and very few, very successful articles. The exact same relationship holds for YouTubers: there are 20 or so channels – like Jake Paul, PewDiePie, who films himself playing computer games, and DudePerfect, who performs trick shots with basketballs, playing cards and other sporting equipment – that have tens of millions of subscribers, while hundreds of thousands of channels have only a handful.

By comparing a model to the double logarithmic plots of citation data, theoretical physicists Young-Ho Eom and Santo Fortunato measured how the relative importance of 'also liking' (citing papers that have already been cited many times) has changed over time. The straight-line power law in 2008 is best explained by a high degree of 'also liking', while a smaller level of inequality seen in earlier years is consistent with a greater tendency of scientists to make independent

decisions. The intervening decades have seen the rise of a
scientific popularity contest, where 'also liked' papers take off
and non liked papers are left behind. The inequality has
continued to accelerate. By 2015, one per cent of papers in
leading journals accounted for 17 per cent of the citations in
those journals.

Given the risk that 'also liking' drives a scientific popularity
contest, we might imagine that our community has been
very careful about how citations are interpreted. It is this part
of the story that is the most ironic. For me, it started in 2005
as a bit of fun. My friend and colleague Stephen Pratt asked
me during a coffee break: 'Have you heard of the h-index?'
No, I hadn't. He explained: 'To have an h-index of h you
should have published h papers each of which has been cited
at least h times.'

Huh? It took me a little while to get my head a round this,
so Stephen showed me my articles as listed on Google Scholar.
I'd only published nine at the time, one that had been cited
seven times, another four times and two more that had been
cited three times each. So my h-index was a lowly three. I
had three papers that had been cited more than three times.
Stephen was ahead of me with an h of six. Once I'd got the
idea, we looked up everyone we knew. Stephen's boss, the
eminent mathematical ecologist Simon Levin, had an h-index
of more than 100: over 100 articles cited over 100 times.

It wasn't long before everyone in academia was talking
about citations and the h-index, and not just during coffee
breaks. Politicians and grant-making bodies quickly caught
on to the idea of using citations to assess scientists. Finally,
they had a way of measuring what was going on inside the
university. For too long, academia had been a closed world,
where taxpayers were expected to trust us to come up with
good ideas. Now, the politicians and administrators thought
they could use citations to measure how many good ideas we
generated.

Around the same time Stephen and I were calculating each
other's h-indexes, I heard the then UK chief scientific advisor
to the government, Lord Robert May, present his article on

'The scientific wealth of nations'.[5] When he took up his advisor role, Lord Bob wanted to find out how the UK compared with other countries in terms of scientific performance. He first calculated how many times UK articles had been cited and the amount of money spent on research. He then divided the first number by the second to show that the UK generated 168 citations per million pounds spent. This was the best research per pound in the world. The US and Canada were in second place with 148 and 121, while Japan, Germany and France all had fewer than 50 citations per million pounds spent. UK science was world-leading, by a significant margin.

This particular conclusion was largely forgotten. Instead, the main message that the UK government and, in subsequent years, governments from other countries, took from Lord Bob's article was that we could now reliably assess scientific success. What ensued was a largely algorithmic monitoring of the state of university departments. Academics were asked to send in a list of their recent articles as part of a research assessment exercise. Since these articles were new, they couldn't be assessed on their number of citations alone: the papers hadn't had time to accumulate citations yet. Instead, the quality of a paper was determined by the 'impact' of the scientific journal it was published in, where impact was assessed by the number of citations of other papers in that journal.

The hunt for impact led, in turn, to an 'also liked' effect at scientific journals. Those journals with a high-impact factor, *i.e.* which published papers that were cited a lot, attracted more and better quality submissions than those with lower impact. Young scientists found themselves competing with each other to get their articles into this small set of highly prestigious journals. Instead of concentrating solely on doing quality work, scientists were scheming to find ways of boosting their h-index and, getting their papers into higher-impact journals.

The 'also liked' effect applies to the scientists themselves. One study showed that articles written by authors who had already written lots of well-cited papers, gained citations more rapidly. It wasn't just the number of citations a paper had, but the reputation of the author that produced success.[6] Monitoring your own or your colleagues' citation records was no longer a bit of fun, as it had been for Stephen and I – it was a necessity for academic survival.

Scientists are a pretty smart bunch of people. When given incentives to produce high-impact research, they do exactly that. To work out the optimal publication strategy, Andrew Higginson from Exeter University and Marcus Munafò from Bristol University drew a mathematical analogy between scientific survival and survival of animals in the wild under natural selection.[7] In Andrew and Marcus's model, scientists can either invest their time exploring new ideas or use it to confirm previous study results. Andrew and Marcus showed that the current research environment, favouring high-impact, encourages scientists to invest most of their time exploring new ideas. Those scientists who perform lots of studies investigating novel possibilities survive, while those who are carefully checking their results 'die out'.

At first sight, this might appear like a good thing: scientists concentrating on researching novel ideas rather than going over the same boring old results. But the problem is, that even the best scientists make honest mistakes. These mistakes include a lot of results that are statistical false positives: if lots of experiments are done in the search for novelty, then by chance, some of these experiments will appear to have yielded an exciting new result, while in actual fact the researchers 'got lucky'.

The luck here is on the part of the researchers who obtained the result. They can now publish their work in a prestigious journal and perhaps obtain a new research grant. For science as a whole, these lucky, but incorrect, discoveries are far from positive. In Andrew and Marcus's model, there is little incentive for other researchers to check these results. Those scientists who were interested in confirming or refuting other people's research have found themselves out of a job after writing low-cited papers. As a result, more and more incorrect ideas accumulate in the scientific canon.

Like all models, Andrew and Marcus' work is a caricature of scientific activity.

Despite the problems, I don't think that citation monitoring and hunting has damaged the quality of the actual research done by scientists. That is, when we get time for research, we scientists still do it well. Most scientists I meet are motivated by the eternal search for truth and the desire to know the right answer. We enjoy finding ways to prove our colleagues wrong by retesting their results. For most of us, contradicting one of our colleague's theories is almost as satisfying as obtaining a new result ourselves. So, incentives remain to test potentially overblown theories and incorrect experimental results.

What the algorithmic assessment of research has done is reduce the amount of time we have to dedicate to pure research. Articles have to be 'sexed up', as we often call it, in order to get into top journals. This involves lifting up the best results and showing why researchers from other fields will be interested in these results. Our work is also 'salami-sliced', another popular phrase we use, into smaller, publishable entities in order to maximise the number of articles we write. All of this sexing up and salami-slicing takes time. We have to write more words and submit and resubmit our papers more times, working our way down from the high-impact journals, which reject most papers, to the lower-impact ones.

It is here that the irony lies. Anurag's Scholar algorithm provided us with a 10 per cent efficiency gain. What did the scientific establishment and the financing bodies do with that efficiency? They used it to monitor and control us. They made us change our way of working so that we lost much of the efficiency we had gained. And they created a rich-get-richer scientific community, dominated by the top one per cent. This suits many scientists, but it leaves other, very good, researchers behind.

Not all scientists have surrendered to the popularity contest. Some have fought back, in the way they know best:

using science. Santo Fortunato has shown that, in the long term, the h-index provides only a very weak indicator of scientific productivity, especially when used to evaluate younger scientists.[8] Of the 25 researchers awarded the Nobel Prize between 2005–15, 14 of them had an h-index of less than 10 when they were 35 years old. An h of 12 is often quoted as a requirement to be hired for a tenure-track position,[9] a rule that would have meant these Nobel laureates would not have been able to find a job.

Albert-László Barabási – who rose to an almost YouTuber-level of academic celebrity early in his career with a paper on 'also liking' and double logarithmic plots – has shown that the most important article that a scientist writes is equally likely to come at any time in their career.[10] It can be the first one they write. It can be an article they write directly after taking their PhD, or when they are working hard to find a permanent position. It can be written when they are established as a researcher, or it can be the last paper they write. The breakthrough can happen at any time. This observation makes it very difficult for grant-awarding bodies to know to whom to award grants, but it suggests that previous citations alone are not the solution. Funding successful researchers might be missing the boat, while neglecting researchers who have worked for years without making a breakthrough might deny us our most important discoveries.

Algorithms such as also liked provide new forms of collective behaviour, new ways for us to interact with each other. These can have many positive effects, allowing us to share our work more rapidly and more widely. But we shouldn't let the algorithm dictate how we see the world. This has happened to a degree in academia. Citation indices and impact, because they were easy to calculate, became the currency of science.

Inequality is one of the biggest challenges facing society, and it is exacerbated by our lives online. We judge each other in terms of the number of friends we have on Facebook, the number of followers we have on Twitter and the number of contacts we have on LinkedIn. These judgements aren't completely wrong: as Duncan Watts and Matthew Salganik saw in their study of music charts in the previous chapter, really bad songs sank to the bottom. But they are not completely right either. The same critique that led me to allege that Jake Paul has only a modicum of talent, can be levelled at successful scientists. Accumulating social capital, in the form of likes and shares, leads to the accumulation of financial capital, be it research grants or Lamborghinis. The feedback then continues.

The also liked algorithm is straightforward to understand, and if you didn't completely get it the first time around, go back to the start of Chapter 9 and reread the algorithm description. You need to understand how it distorts inputs and outputs because it is definitely operating in some aspect of your life. Whether you are building up LinkedIn contacts to impress employers, or you are an employer comparing a brash candidate with a wide social network to a quieter candidate with very few friends on Facebook, you should not let the algorithm alone decide things for you. Our human integrity is one of the most important things we have.

There are other ways we could organise our lives online.

The 'also liked' algorithm is not the only way of sharing information. I asked Kristina Lerman which 'sharing' service she thought had the best system. She went for the original version of Twitter. Before 2016, Twitter simply showed the things shared by the people you followed in time order. If you weren't on the site when your friend posted something, the chances were that you would miss it. Retweets by other users could help, but time was the main factor that determined what we saw.

Gradually, Twitter has moved towards an 'also liked' algorithm. It has an 'in case you missed it' feature that lifts heavily liked and retweeted content higher in your timeline. The default option in your settings is now to 'show me the best tweets first'. Turn this off. Expose yourself to as wide a variety of opinions as possible.

One app that has minimal filtering is Tinder. I don't have Tinder and, while I am prepared to dissect much of my online life in order to find out how algorithms work, I draw a line at downloading an online-dating app. My wife, understanding as she is, would not condone me creating an account, even if I claimed it was for scientific purposes.

Lots of my younger colleagues have been keen to explain Tinder to me, occupying as it does large portions of their time. As a user, you are shown a series of profile pictures of people you might be interested in and who are located nearby. Swipe right if you like what you see, swipe left if you aren't interested. If you make a right swipe for someone and they make a right swipe for you, then you can chat with each other through the app and hopefully romance (or whatever it is you happen to be looking for) will blossom.

With the emphasis on the picture, users' profiles contain just a short bio, including your name, age, interests and a short bit of text about yourself. When it was released, Tinder was an antidote to the other online-dating sites that boasted complex algorithms aimed at finding your perfect match. Tired of filling in questionnaires or having their Facebook profiles analysed, young people appreciated its simplicity and honesty. Unfiltered, Tinder allows you to assess what is on offer for yourself.

There is a massive difference in how men and women swipe. In one study conducted in London using fake accounts and a single profile picture, females were about 1,000 times more likely to receive a right swipe from a man than males were to receive a right swipe from a woman.[11]

If you are one of those lonely men who are not getting any right swipes, there are a few things you can do. Firstly, writing a bio quadruples the probability of being chosen, as does including two additional profile pictures. Women want more information and are choosier than men, so if you want to get a match then you need to give them what they want.

These basic tips don't completely solve the problems that men are having. Gareth Tyson, who conducted the London study, also sent out a user questionnaire to find out how many matches (right swipes by both parties) people got. The majority of men got matches for less than 10 per cent of their right swipes. A lucky five per cent got more than 50 per cent matches. Although the algorithm is very different, the inequality is very similar to that seen between scientific papers: a very small group of probably very good-looking guys get all the attention. The rest of you men have to do a lot of swiping before you get a match.

Or do you?

Alex, my colleague with whom I analysed FiveThirtyEight, has a lot of first-hand experience of Tinder. He came to Sweden as a young single man from Australia and, at first, the dating app seemed an ideal way to meet new people. But he soon found that he was one of the men who wasn't getting dates and wondered what he might be doing wrong. His answer was to create a mathematical model of how both men and women behave on the site. When you aren't getting any dates you have more time for maths, I suppose.

Alex realised that there was a feedback between failing to get a match and swiping more. When men first use Tinder, they tend to be quite picky, but when they find that they aren't getting any matches, they widen the search and right swipe even more. Women do the opposite. Since they are getting too many matches, they narrow down the search. Alex's model showed that this feedback made the situation worse for everyone. In the end, women pick only one or two men and men pick almost every woman. Alex called it the 'Unstable Dating Game' because the stable solution of finding perfect pairings disappeared for both men and women.[12] Instead, the outcome of the algorithm was the same as the one Gareth Tyson observed in his study of London users: on average, men right-swiped about half of the women they were presented with, while most women right-swiped less than 10 per cent of men.

Alex has found a solution that works for him. He went in the opposite direction of the herd. He decided to become both more patient and more demanding in his choice of dates. By carefully targeting the women he really believed in, writing a bio that he thought they would find interesting and then waiting until some of them started to choose him too, his match-rate dramatically increased. He hasn't found true love on Tinder yet, but he has had many enjoyable meet-ups in Stockholm's cafés, and even started a band with one of his dates.

An 'also liked' system for dating wouldn't really work. 'Yes, he was brilliant. Maybe you should try him?', is not a notion we often hear shared among friends, online or otherwise. But a swiping system for academic articles is an idea worth pursuing. Each morning, when I come into the office, I could sit down and swipe my colleagues' papers left and right without an algorithm telling me what my peers have cited. If we right swiped each other, this app could put me in touch (one-to-one without anyone else knowing we are friends) with colleagues so we could chat a bit more about the science.

Anurag Acharya, if you are reading, maybe you could get on to it. And finally ... I'll be able to swipe without destroying my marriage.

CHAPTER ELEVEN

Bubbling Up

The UK voting to leave Europe and the election of Donald Trump were big surprises for those of us living sheltered lives in academia. Most of my liberal-minded colleagues didn't understand what was going on. They didn't know anyone who would vote for Trump, and had never met anyone who wanted to split up the European Union. The liberal newspapers they read were equally taken aback, running articles with titles like: 'Meet 10 Britons who voted to leave the EU,' and 'Why the white working-class voted for Trump'. There was a desperate need to explain how and why voters suddenly didn't agree with what many of us believed to be an established consensus about how things should be done.

I was trying to make sense of the vast number of articles that had been written on the politics of the previous year. One prominent explanation was that algorithms were to blame for misinforming people. The *New York Times*, the *Washington Post*, the *Guardian* and the *Economist* were among a large number of media outlets that ran mathsy-sounding stories about the isolation and polarisation created by algorithms.

To start with, the discussion was of echo chambers and filter bubbles. The theory was that Facebook and Google were personalising our searches to such an extent that we only saw what we wanted to see. The focus then turned to fake news. Teenagers from Macedonia were automatically generating news stories, trying out different combinations of nonsense rumours about Trump and Clinton, in order to create traffic for their websites and advertising revenue. It was claimed that Russia was influencing the election using Facebook adverts and paid Internet trolls, who would argue aggressively on Twitter and various political blogs in order to create polarisation.

A lot of the concern over maths and algorithms resonated with what I had already found out about 'also liked' and the algorithms Facebook used to analyse our personalities. It appeared as if we were letting algorithms tell us what to think and do. There was a genuine risk that some of the news presented to us was skewed or made up by politicised black hats. But I felt uneasy with the way maths was used in these stories, and what they implied about the people consuming the media. Were the American people really so stupid that the only messages they had failed to filter out were those generated by Macedonian teenagers and Russian trolls? Were people really so strongly influenced by what they saw on Facebook? A lot of my colleagues seemed to think so. I wasn't so sure.

Long before academics and journalists were worrying about the Trump and Clinton filter bubbles, two young computer scientists were already looking at how political campaigns shaped and were shaped by the Internet. In 2004, Lada Adamic and Natalie Glance studied the 'blogosphere' in the run-up to that year's presidential election.

In comparison with today's social media, these blogs appear rather quaint. The format was simple: text explaining the blogger's point of view, a few pictures pasted in from newspapers and links referencing news websites and other blogs. There weren't 'like' and 'share' buttons connecting to social media. Facebook hadn't taken off and Twitter was yet to be invented. Instead, political blogs connected via direct links between web pages, often in the form of 'blogrolls' listing sites and articles the blog's author approved of.

Figure 11.1 shows the links between the top 20 liberal blogs (sympathetic to Democrat party, black circles) and top 20 conservative blogs (sympathetic to Republican party, grey circles) in the run-up to the 2004 election. The size of the circle representing each blog is proportional to its popularity among other blogs. The line thickness indicates the number of links between the blogs.

The 2004 blogosphere was divided. Democrat blogs linked nearly exclusively to Democrat blogs. Republican blogs linked nearly exclusively to Republican blogs. There were

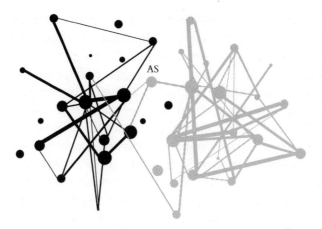

Figure 11.1 Network of the top 40 political blog connections in run-up to the 2004 presidential election. Black circles indicate liberal blogs. Grey circles indicate conservative blogs. The area of the circles is proportional to the number of times a blog is linked to by other top 40 blogs. The thickness of the lines is proportional to the number of times one of the blogs is linked to another, with only blogs with five or more links shown. The blog labelled AS is The Daily Dish, *run by Andrew Sullivan. Drawn based on data from the paper: 'The political blogosphere and the 2004 US election: divided they blog', by Adamic and Glance (2005).*

very few links between the two worlds. The only Republican blog with more than a few links to a Democrat blog was The Daily Dish, run by Andrew Sullivan (marked AS in Figure 11.1), and he had switched his allegiance for the 2004 campaign to support Democrat John Kerry. Otherwise, the segregation was pretty much complete.

Democrat and Republican blog networks were different from each other. If you look closely at Figure 11.1, you'll see that conservative blogs have more connecting links than liberal blogs. Conservative bloggers tended to comment more on each other's writing than liberals did. There was a more active internal discussion between Republican sympathisers. Importantly, however, neither the liberal nor the conservative networks were closed to the outside world. About every other post, from both political persuasions, contained a reference to a news article in the mainstream media. For example, the *Washington Post* was cited around 900 times by liberal blogs and 500 times by conservatives during the two and half months leading up to the election. Liberal and conservative blogs were segregated, but they both let the mainstream media in.

Lada and Natalie's study was a hint as to how computer scientists would analyse news and politics in the future. The methods they developed during their study – for dissecting the political network, automatically identifying the keywords within debates and finding out how political commentators connected – were changing the way we understood the media. In the article, they wrote that in the future, 'we would like to track the spread of news and ideas through the communities, and identify whether the linking patterns in the network affect the speed and range of the spread'.

In an academic context, there were now many exciting possibilities. Lada and Natalie's study showed that it was possible to use mathematical methods to understand how we communicated about politics. But if scientists can develop methods for understanding political communication, then political parties and large companies can use these methods to manipulate how *we* talk to each other. And from 2004 onwards, those methods developed extremely rapidly.

By the time of the 2016 US election and Brexit vote, Breitbart and the Huffington Post had become the right and left-wing amalgamations of these early political blogs. Many of the 2004 bloggers were working for these websites, or others, such as the Drudge Report. Some, like Andrew Sullivan, were writing articles about how they had burnt out from their constant social media activities. But there were plenty of new voices to take their place: independent political blogs pop up all the time, and tens of thousands of people use online publishing platform Medium to document their every thought. Articles are 'upvoted' and 'downvoted' on Reddit; they are spread on trending sites like BuzzFeed and Business

Insider; they are aggregated on Feedly and Fark; they are turned into spontaneous newspapers on Flipboard, and they are shared on Facebook. All of these millions of words, written every day from every corner of the world, are then commented on, discussed, ridiculed and/or supported using the 280 characters provided by Twitter.

When commentators analyse this vast array of social media they usually come back to two key themes: the echo chamber and the filter bubble. These concepts are related but slightly different. The 2004 political blogs network is a primitive example of an echo chamber. Bloggers linked to other bloggers who agreed with them, confirming their views and supporting what they already thought. If you clicked from blog to blog, choosing a link at random from each page you came across, you would remain stuck within the same set of opinions you started with. If you started on a liberal blog page in 2004, then the probability that 20 clicks later you would still be on a liberal page is over 99 per cent. Start on a conservative page then 20 clicks later you will probably still be reading conservative material. Each set of bloggers had created their own world, within which their views reverberated.

Filter bubbles came later and are still developing. The difference between 'filtered' and 'echoed' cavities lies in whether they are created by algorithms or by people. While the bloggers chose the links to different blogs, algorithms based on our likes, our web searches and our browsing history do not involve an active choice on our part. It is these algorithms that can potentially create a filter bubble.[1] Each action you make in your web browser is used to decide what to show you next.

Whenever you share an article from, for example, the *Guardian* newspaper, Facebook updates its databases to reflect the fact that you are interested in the *Guardian*. Similarly, when you share from the *Telegraph* the database stores an interest in that newspaper. In a talk to publishers in April 2016, Facebook's Adam Mosseri, who is responsible for the development of the news feed algorithm, explained that 'when you first sign up for Facebook, your news feed is a blank slate. Then slowly but surely over time, you friend the people you care about and follow the publishers you are interested in. And you build your own personalised experience'.[2]

Facebook's algorithm decides what information to show us on the basis of the choices we have already made. To understand how its filter works, imagine that you are a relatively open-minded and independent person who has just signed up for Facebook. Let's assume that you are so open-minded that you read both the right-leaning *Telegraph* and the left-leaning *Guardian* newspapers, and you share articles from both of these publications. Let's also imagine that you have left-and right-leaning friends in roughly equal numbers. I know this is unlikely to be true in a literal sense, but in order to see the potential effect of the algorithm, we start off by assuming that you are the most open-minded individual possible.

Now you start posting. Imagine you make a few posts, sharing articles from both the *Guardian* and the *Telegraph*. Your friends don't take much notice at first, but then one of them comments on a link you share from the *Telegraph* about problems with corruption within the European Union. You reply, and the two of you write backwards and forwards about the article, liking each other's posts. Now you are giving Facebook what it wants: information about what makes you spend time on its site. And in return, Facebook can give you back more of what it thinks you want. The next day, a post by your friend, moaning about a new EU regulation, appears at the top your news feed. The next article down is also from the *Telegraph*; another friend has shared information about the business advantages for a post-Brexit UK. Both these articles catch your attention and you start to comment. Facebook notes your further interest and the next day it provides a few more EU-critical articles. The filter slowly forms around you.

This description is just a single story, but we can use some mathematics to understand how the filter algorithm will typically operate. Facebook determines how visible a recently shared newspaper article is on your news feed, based partly on the following equation:[3]

Visibility = (your interest in newspaper) × (closeness to friend sharing article)

When you interacted with your friend about the post you shared, you boosted both of the factors in this equation: you showed an increased interest in the *Telegraph* and caused Facebook to increase its measure of closeness between you and your friend. So we can think of visibility increasing as engagement squared, *i.e.* vibility is proportional to engagement through interest in the newspaper multiplied by engagement through closeness to friend. As a result, the visibility of future *Telegraph* articles increases. Increased visibility makes it more likely that you click on these links in the future, which further increases Facebook's ranking and leads to even more visibility.

The next step is to formulate this as a mathematical model that captures both the algorithm's behaviour and a user's interaction with the algorithm. I now describe this in the form of what I call the 'filter' model. As in my 'also liked' model of Amazon, the 'filter' model is a simplification of how Facebook's algorithm actually operates. It captures the most central feature by which Facebook filters our feed, Twitter filters our timeline and Google filters our searches: the more we click on something, or someone, the more prominently they are shown to us, and the more likely we are to keep clicking on them.

The 'filter' model works over a number of interactions. On each interaction, a user is shown posts from two sources; following on from my earlier example we'll call them the *Guardian* and the *Telegraph*. Assuming that the visibility of these two newspapers is determined by the visibility equation above, and that the users click on a newspaper in proportion to its visibility, we can simulate how the relative visibility of the two newspapers changes over time.[4]

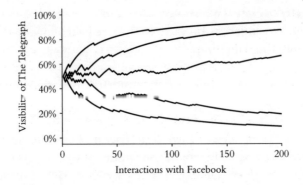

Figure 11.2 Simulation of how the 'filter' model works for five different 'unbiased' users. On each round of the model, the user chooses whether to interact with a post from the Guardian *or from the* Telegraph. *If they choose the* Guardian *then the visibility of the* Guardian *articles increases, as does the probability that on the next round they again choose the* Guardian. *This feedback leads users to eventually have increased visibility for one of the two newspapers over the other.*

Figure 11.2 shows how the visibility of *Guardian* and *Telegraph* articles changes for five different simulated users, all of which started with 50 per cent visibility for both newspapers. After 200 interactions with Facebook, two of the simulated users of the 'filter' algorithm have a high level of visibility for the *Guardian*, two have high visibility for the *Telegraph* and the last has a slightly higher visibility of the *Telegraph*. Simulating more users produces similar results: after 200 rounds of filtering, most users are shown more posts by one of the two newspapers.

Remember, these simulated users had no initial preference for either newspaper. The preference forms as the clicks change the visibility. The feedback between the choices they have made and the posts they are shown determines which newspaper they are ultimately shown more often.

As Adam Mosseri says, to start with 'you are a blank slate'. But as soon as the users write their first message on that slate, the feedback between visibility and engagement begins. Because visibility is proportional to engagement multiplied by itself, the slate is rapidly filled with the newspaper that, by chance, the user first showed an interest in. The 'filter' algorithm creates a bubble, even for initially unbiased people.

For people who begin using Facebook with an established preference for either left or right-leaning media, the effect is even stronger. The 'filter' algorithm picks up small initial differences and exaggerates them until the other side of the argument is lost. The users are locked in to self-confirming ideas and interactions with a smaller group of friends.

The algorithm Facebook applies to your feed is a bit more complicated than my 'filter' model. It claims to use over 100,000 personalised factors to make decisions about what to show you (by which it actually means it has done a principal component analysis of your 'likes'). So while the model shows that there is a risk that Facebook creates filter bubbles, it does not prove that all its users are trapped in bubbles. I wanted to find out how well the simplified filter model captures reality.

Mathematician Michela Del Vicario is one of a group of researchers at the Laboratory of Computational Social Science in Lucca, Italy, who is putting Facebook to the test. The researchers identified 34 Italian Facebook pages that share scientific advances and 39 that share conspiracy theories.[5] They studied how Facebook users shared, liked and commented on the posts made on these pages. There was lots of evidence of polarisation between these two communities. People who 'like' and share science seldom 'like' and share conspiracies, and vice versa. There was also evidence of an echo chamber within each community. Most conspiracies spread primarily between people who are already 'liking' and sharing conspiracy theories, and make little impact on the Italian population as a whole.

When I spoke to Michela, she described a vicious cycle: 'The more conspiracy articles a person has shared, the more likely they are to share more of these stories and the more likely they are to interact with other people interested in conspiracy theories.' This is the process described in my filter model: there is a feedback between sharing conspiracies and increased exposure and sharing of similar stories.

Depressingly, the same is true for science. A small bunch of Italian science geeks share the latest science news with each other, while the wider general public pays little or no attention to them.

Michela and her colleagues also analysed the words used in conspiracy and science posts. She told me: 'Apart from a few very active users who are quite positive when they post, the general rule is that the higher the activity of the user, the more negative words they use.'

While both scientists and conspiracy theorists tended to be less positive the more they posted, the effect was stronger for scientists. The scientists used fewer positive words than conspiracy theorists, and the more active they were on Facebook, the more negative they became. Becoming a dedicated member of an echo chamber is not a route to happiness.

Not only are conspiracy theorists less grumpy than scientists, their shared posts are also more popular than those made about science news. This is particularly worrying since many of the conspiracy theories are *about* science. One of the continuing rumours is about a link between vaccines and autism. These persist despite repeated rigorous scientific studies, shared by the scientists, that show there is no connection. Another prominent and ongoing theory is the so-called 'chemtrails conspiracy': that governments are seeding clouds with toxic chemicals and infections using aeroplanes.

Watching chemtrails videos, I am genuinely surprised – not so much about the content of the films, but about my own reaction to them. Sitting alone at home, in front of my screen, on a quiet evening after work, I can feel myself slowly starting to believe what I am watching. A video with 580,842 views on YouTube shows a meeting in California. Testimonials of pilots, medical doctors, engineers and scientists are clipped together. The setting is a local government hearing, the room is packed and the impression is of an important investigation underway. Grey-haired men step up, one after the other, and give authoritative quotes on the 'pervasive nature of nanoparticles', 'increased air pollution' and 'dramatic drops in insect species'. They talk about Alzheimer's disease, autism, ADHD, ecosystem collapse and pollution in rivers. Some of what they say makes scientific sense, focusing on water quality and environmental problems, and I feel myself being sucked into their arguments. The film moves rapidly from one speaker to the next, and the camera repeatedly pans out to an audience waving hands in support of the speakers. I have questions I want to ask. But everything moves so fast, and I don't know the context for each of the statements. I can't quite put my finger on what is wrong.

This is the effect of a well-made scientific conspiracy film. I am overwhelmed by the juxtaposition of fact and fiction, trying to make sense of the jumble of ideas. I click from one video to the next, and end up spending over three hours browsing videos, each of which has been watched hundreds of thousands or even millions of times before. I watch a stern-faced presentation from a member of anti-chemtrails organisation Geoengineering Watch, about 'the rapidly-evolving science' behind chemtrails. I find a clip of Prince talking about how the link between chemtrails and violence inspired his songs. There is a concerned mother calmly explaining the connections between heavy metals and human health. These are rounded off by confessions from retired government officials and a woman who reportedly had her children taken away from her because she dared to reveal the truth about how our government is poisoning us. Afterwards, I shut my laptop and sit there in the darkness. It takes a few minutes before my mind clears and my scientific brain starts to work again.

I don't think I am at serious risk of being sucked into a conspiracy bubble, but watching the videos allows me to better understand those people who are. I go on to look at the dry, factual webpage curated by Harvard professor of Geoengineering, David Keith, where he carefully explains why the chemtrails conspiracy is entirely fictional. I watch a video abstract recorded by Steve J. Davis, Earth System Science professor at the University of California, who surveyed 77 scientists about the possibility of a chemtrails-like conspiracy. All but one of these experts said that the so-called evidence could be explained by other wellunderstood factors.

But Steve's video had only been watched 1,720 times and I could see why. It was factual, but it lacked drama. It showed him sitting casually in his office, talking in a neutral voice about the importance of peer review. Without the agenda of an organisation like Geoengineering Watch, he didn't feel he needed to push his side of the argument. I could understand why he presented his work in the way he did, but comparing his videos with those I watched earlier, I could also see why the scientific bubble is smaller than the conspiracy bubble. Conspiracies are much more captivating.

Within conspiracy chambers, ideas go unchallenged. The same people continue to share the same material with each other. Mike Wood, lecturer at the University of Winchester and writer for the blog 'The Psychology of Conspiracy Theories', told me that 'within established communities, conspiracy videos often have dissenting comments buried'. Many YouTube conspiracy channels have disabled comments, others are filled with messages explaining that those who make comments contradicting the conspiracy, are themselves part of the conspiracy. The anti-conspirators are said to be proof that the conspiracy exists.

Mike enjoys arguing about conspiracy theories on the Internet. His blog has an extensive comments section where he sincerely answers questions, however daft or uninformed these happen to be. Despite Mike's patient tone, this comments section sometimes become a forum for conspiracy theorists and anti-conspiracy enlighteners to throw insults at each other. The section about the predictive programming conspiracy, had filled with comments by two interlocutors accusing each other of 'disingenuous crap', believing 'alien bullshit', having a 'paranoid world view' and finally 'plucking turds out of your own ass'. Mike made no further comment.

According to Michela Del Vicario, the aggressive debates only serve to fuel the conspiracy theorists. She found that the more negative comments that a conspiracy theorist encounters, the more likely they are to continue to share, comment and argue over them. Aggressive attacks from the outside only act to reinforce the bubble.

There is one conclusion from both Michela and Mike's work that offers a glimmer of hope in the battle against conspiracy theories. Once conspiracy theories start to spread, they meet with a wider resistance, both from people typically interested in science and from members of the general public. The comments under the most-watched chemtrail videos contained brief explanations of the science ('that is a video of a plane dumping fuel'), reasonable arguments as to why the theory was implausible ('if the government are really poisoning you, why do they do it with opaque trails of gas when they could just poison you with chemicals you can't see?'), along with the usual array of patronising insults ('I hope you people can't vote nor multiply' and 'MORONS!!! You people are idiots').

These comments ensure that the wider general public, those outside the conspiracy bubble, can easily see that a particular Facebook post or YouTube video was controversial. Mike told me: 'If there's just one dissenter they'll often be ignored, or will end up arguing 50 comments deep with a single person, but if the video gets some outside attention, the dissenting comments start to get 'thumbs-up' likes from the outsiders.' My own limited survey of conspiracy videos – I was now into my fourth evening in a row of compulsive viewing – concurred with their results. I have even started 'liking' the dissenters and 'disliking' the supporters. I took Michela's advice though, and resisted adding sarcastic comments of my own.

Most conspiracy theories are kept inside a bubble, where the theorists support each other's ideas. If the bubble expands too much, then the non-conspirators could burst it. Bigger echo chambers provide room for more dissenting voices.

The only people who have access to enough Facebook data to fully dissect a large-scale political debate, are the scientists who work there. After publishing her paper on the political blogosphere, Lada Adamic went on to become a leading researcher in the field of computational social science. She worked as a professor of Computer Science at the University of Michigan, publishing some of the most influential papers in the mathematical social sciences. Then in 2013 she took a sabbatical to work at Facebook, where she ended up staying as a data scientist.

Working at Facebook allowed Lada to test the filter bubble hypothesis on mainstream politics. Together with two other Facebook scientists, Lada looked at the network of connections between politically affiliated friends on Facebook. Friendship networks on Facebook were not nearly as politically segregated as the 2004 political blogs: 20 per cent of the friends of liberals were conservatives, 18 per cent were moderate and 62 per cent were liberals. So while there was certainly a tendency by both liberals and conservatives to prefer the friendship of like-minded people, both political persuasions are exposed to a good proportion of people who did not agree with their views.

The Facebook researchers then looked at the news content shared by friends of conservatives and liberals. Of the content shared by friends of conservatives, about 34 per cent was liberal news. If we compare this with the average amount of liberal news shared on Facebook, which is 40 per cent, then we see that the bias towards liberal news is small. Facebook friends provide a faint hum of agreement, rather than a chamber full of echoed opinions. Friends of liberals were more likely to share news that reflected these views. Liberals' friends shared 23 per cent conservative content, compared with a Facebook-wide average of 45 per cent. Here the echo was audible, but far from an overwhelming din.

Most of us recognise this varied picture of Facebook friendships and what they share. We have a lot of friends from school, many of whom we weren't even really friends with at the time. We have friends through work and from the places we have been on holiday, studied and lived. I have probably travelled a bit more than the average Facebook user and now live abroad, but in most aspects, I'm a pretty typical British male. The median number of Facebook friends for a user is 200.[6] I have 191 friends. They come from a wide range of backgrounds and, although they are typically more left-leaning liberals, they have a wide range of political opinions.

We know that Facebook does not treat our friends equally. The visibility/closeness equation (see page 139) means that those friends who I interact with most are shown on my timeline. I haven't interacted much with some of my old school friends and the Facebook algorithm knows this. It doesn't show them as prominently on my timeline. Lada, and the other Facebook researchers, wanted to find out how this filtering affected the political views we saw. If the 'filter' model applies to political news sharing, then we would expect the Facebook news feed to provide less exposure to contradictory ideas than articles shared by friends.

Their results were very clear. The effect of filtering was negligible. Both liberals and conservatives were exposed to only slightly less opposing views through the filter, than if Facebook had provided random posts on their feed. We *are* more likely to see articles on our feed from our close friends, but the political views expressed are *not* more extreme than those expressed by all our friends. A large proportion of what we see on Facebook does not conform to our own opinions. Moreover, American conservatives, a group often accused of being closed-minded, were exposed to slightly more contradictory opinions than liberals.

The filter bubble test was published in the leading scientific journal *Science*. It was just one example of how, during the first half of this decade, Facebook took science seriously. It employed leading researchers to find out the effect of its product – and these researchers thought big.

In the run-up to the 2010 congressional election, 60 million Facebook users in the US were shown a message, placed at the top of their news feed, that helped them find their polling station and provided them with a button to indicate they had voted. The message contained pictures of the users' friends who had already pressed the 'I voted' button.

James Fowler, professor at the University of California, decided to measure the effect of this message. Working together with Facebook, he and his colleagues created a smaller group of users, of around 600,000 people, to whom they showed the same message without the faces of friends who had voted.' Their hypothesis: it is the social nature of messages that creates an effect on users. By seeing the faces of our friends who had voted and giving us a chance to tell them that we are with them, we are encouraged to go out and vote.

The hypothesis was correct. The users who saw the non-social message were slightly less likely to vote than the users who saw which of their friends had voted. The difference in voting probability was only 0.34 per cent. But even a small change in voting behaviour can make a massive difference to the number of people voting. By matching Facebook users against the electoral register, James and his colleagues estimated that they had created at least 60,000 new voters as a direct result of having seen the message, and at least a further 280,000 who had voted as a result of the social contagion created by the message. A small nudge made a big change to the number of people participating in democracy.

The political study showed that Facebook messages could have an effect on our everyday lives. The next step, taken by Facebook data scientist Adam Kramer, together with two Cornell researchers, was to look at the overall emotional effect of messages. They conducted a full-out experiment. The researchers removed between 10 per cent and 90 per cent of positive posts from the news feed of around 115,000 users, providing them with more negative posts than usual. Happy posts of families having fun together or amusing pets were temporarily removed from these users' news feed. The researchers then measured the posting behaviour of these users in comparison with a control group who received their usual news feed. The scientific question was how the removal of positive posts would affect the posts that the users, who had been subjected to the treatment, went on to make.

At the time this article was published, concerns were raised by members of the scientific community and a massive outcry was produced in the media about the ethics around the experiment. The objection was that Facebook was manipulating users' feeds and 'deceiving' them without their consent. This could, under certain regulatory guidelines, be considered to be unethical. But the debate in the media and online was less concerned about the details of which ethical rule applied to the particular experiment, and much more concerned about the fact that Facebook was manipulating us.

I found this particular outcry slightly confusing. Of course, Facebook is manipulating the information its users have access to! That is the idea that its business model is built around. News feed manipulation is the algorithm Facebook's Adam Mosseri proudly presents to business leaders. The company is continually adjusting what it shows to you in order to make you use its site more often. It is completely open about this point and even allows you to fine-tune your experience.

The remarkable result, from Adam Kramer's experiment, was that manipulating the news feed had *almost no effect* on users' emotions. The study found that the percentage of positive words used by people whose positive news was removed from their feed was just below 5.15 per cent, while in the control group, with a standard feed, it was near to 5.25 per cent. That is a difference of 0.1 per cent. To get an idea of how small a 0.1 per cent effect is in terms of Facebook posts, imagine you are a relatively active user who posts about 100 words per day. A 0.1 per cent decrease implies that during the next 10 days you will write a grand total of one positive word fewer. Maybe next Wednesday you'll describe a film you saw as 'OK' instead of 'good'.

The results were even weaker when it came to negative words. The effect size was 0.04 per cent, corresponding to roughly one extra negative word per month. It is hard to imagine that anyone else would notice the additional 'boring' you added to a description of a work meeting, or the fact that you were 'disappointed' your football team lost a qualifying match.

While newspapers were debating the dangers of Facebook experimenting on us, they ignored the fact that the results were negligible. This was partly Adam Kramer and his colleagues' fault. The title of their paper was: 'Experimental evidence of massive-scale emotional contagion through social networks'. What?! In hindsight, this title, which implies that emotions spread like Ebola through our social networks, was not a particularly accurate description. What they had actually found could be better described as a tiny but statistically significant increase in the expression of positive emotion as a result of a large-scale manipulation of Facebook users' news feed.

There is a lot of hype about Facebook and its effect on our lives. But what struck me, after carefully reanalysing the big studies performed and talking to the researchers involved, was that the results were nearly always distorted or exaggerated when reported in the media.

The hype jarred with my own understanding of the science. Yes, Facebook could blow up a small bubble on election day that got a few more people out to vote. Yes, it could ever so slightly deflate our emotional bubble by only showing us depressing posts. And yes, Facebook doesn't provide news that is entirely representative of the wide variety of opinions expressed throughout the world. But none of these were life-changing effects. The effect of Facebook on our life is very weak compared with the effect of our everyday interactions with people in real life.

I could see that that the bubble analogy was useful. Michela and her colleagues in Italy showed how insular certain groups could become. But the weakness of the bubble theory, when applied to larger groups of people with wider interests, was starting to bother me. Social media was full of algorithmically-created bubbles, and there were some really crazy conspiracy theorists floating around in them. But for most of us, there seemed to be a way out of the bubbles. What was it that prevented us from becoming trapped? If Facebook's 'filter' algorithm is capable of locking us into a particular viewpoint in theory, how do we escape in reality?

Why are our emotions largely unaffected during the long periods of time we spend using social media?

The only way for me to answer these questions was to climb inside my own online bubble and see if I could find my own way out. It wasn't something I was particularly keen on doing. I have a bubble that I am very fond of and don't particularly want to burst. But I couldn't make excuses any longer. I had to find out what my own bubble was made of.

CHAPTER TWELVE

Football Matters

My use of social media revolves around football. I have a Twitter account, @Soccermatics, where I talk to fellow maths-football nerds about the game, and the maths and stats used to analyse it. This small part of the Twitterverse is both a filter bubble and an echo chamber. I know that Twitter filters my feed so that the biggest geeks are at the top of my home page. Other football stats accounts, such as @deepxg, @MC_of_A and @BassTunedToRed all say things I am interested in, and Twitter makes sure I see their tweets first. My tendency to follow other like-minded geeks creates a bubble where we all agree about the need for maths and football to work together.

Occasionally, less nerdy Twitter users (usually Chelsea fans) who don't appreciate my latest passing network visualisation, pop into my echo chamber and tell me I should, to choose a few representative examples, 'go and wank over your spreadsheet', or 'meet girls instead, you fucking virgin'. But, since these statements don't contain much scientific information, I usually ignore them or politely link the Chelsea fan to an article explaining the basic concepts behind football analytics.

I freely admit to living in a football analytics bubble. But I have noticed an interesting thing about this bubble. Some of us, like me, are Liverpool fans. Some support Everton. Some support Manchester United and others City. Some of us even support Chelsea. There are Real Madrid fans, Barcelona fans, fans of Italian and German football. And there are analysts who tweet about African football, about Asian football and soccer in the US. Instead of residing in a bubble for each team or even each league, us football maths geeks see and share information about all forms of football. I know just as much about Manchester City now as I do about Liverpool. I have learnt about Major League Soccer in the US (and Canada, as I found out recently) and even more about the women's league, NWSL. I have also peeked outside the soccer bubble to the American football bubble, the ice hockey bubble and the basketball bubble.

Politics also seeps into my bubble. There is often a strong working-class, socialist ethos among many of the Liverpool fans I follow. But I also follow journalists at right-wing UK newspapers and media outlets. Many of the analysts I follow from the US tend to support the Democrats and retweet anti-Trump articles, but I also follow coaches working at a grassroots level across the country, who sympathise with Republican ideals and who are motivated by Christian beliefs. One analyst in particular, @SaturdayOnCouch, keeps me entertained with both his heat maps of German football and his friendly(ish) trolling of anti-Trump liberal football analysts.

I see calls for Manchester United to start a women's football team, campaigns by Leicester City fans to stop homophobic abuse on the terraces and pleas by Borussia Dortmund fans to welcome refugees from Syria to Germany. I see political campaigning and stories about how dissatisfaction among working-class Americans led to the rise of Donald Trump.

The imagery of politics seeping into our social media bubble is a useful way to visualise how we come across different political viewpoints. Bob Huckfeldt, professor at the University of California, Davis, has spent more than 30 years investigating our political discussions. During the 1992 Bush–Clinton campaign, Bob and his colleagues asked supporters of both parties to list the political affiliations of the people they had talked about politics with. They reported that 39 per cent of political discussions were with people who did not support the same presidential candidate as they did.[1] This statistic could be partly accounted for by voters better remembering adversarial discussions, but not entirely. In the run-up to the Bush–Clinton election, people all over the US were regularly discussing politics with people who opposed their point of view.

Bob emphasises that people 'carry out their lives within various contexts, defined along multiple dimensions'.[2] Sports and politics are an example of two different social dimensions. At the game, and in online discussions afterwards, sports fans meet people with all sorts of different political views. The same is true for music, books, films, food and celebrity gossip. People with very different political backgrounds meet each other online to talk about *The Girl on the Train* or *The Avengers*, and end up discussing politics.

Does the 1992 campaign belong to a unique golden age of friendly work colleagues and sports fans chatting over lunch or in the bar about the pros and cons of two political candidates? Not at all. Bob has observed similar levels of discussions that bridge the political divide in Germany and Japan. He also found the same result in what was, until recently, considered the most contentious American presidential campaign in the past 50 years: George W. Bush versus Al Gore in 2000. Over a third of political discussions between the voters surveyed during that election, were between supporters of opposing parties.

These political discussions are not just, 'Booya, I know better than you'. Bob and his colleagues also tested the political knowledge of the people in their studies, and found that it was the experts that others turned to when they wanted to find out more about an issue. Irrespective of their views, those people who knew more about politics were more likely to be asked for their opinion.

The question remains, however, as to whether or not social media has – in the 18 years since Bush vs Gore – transformed how we exchange information about political issues. Bob's research was done in an age before the arrival of Twitter, Facebook and Reddit. Recently, there has been a lot of speculation that things really are different now.[3] Given the concerns about online polarisation during campaigns like the Brexit vote and the US presidential election, I wanted to see if Bob's work still applied today or if echo chambers had taken over.

I decided to test this idea, starting on myself.

Figure 12.1 shows part of my social network on Twitter. Each circle in the figure is a person who I follow and who follows me back. I am a circle in the middle with all the lines connecting to me (since I am the focal point of my own network). Other lines drawn between pairs of people indicate that they also have this follow/follow back relationship. This figure only includes 'typical' Twitter users, which I define as people who follow fewer than 1,000 people and are followed by fewer than 1,000 people. Academic hotshots, celebrities and media personalities are thus excluded.

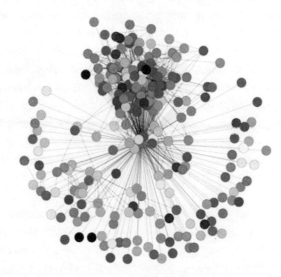

*Figure 12.1 My Twitter social network. All Twitter users (with less than 1,000 followers) who follow me and I follow back are represented as circles. I am a circle in the middle. The lines connect people who follow each other. People with very dark shading have a low degree of separation from the three UK newspapers that published more articles with a 'pro-remain' slant in the run-up to the Brexit vote (*Daily Mirror, the *Guardian and* Financial Times*). A very light shading indicates a low degree of separation to newspapers with a more 'pro-leave' slant (*The Times, Daily Star, the *Telegraph, The Sun, Daily Mail and* Daily Express*).[4] The grey shadings indicate a 'degree of separation' between these two extremes. Figure created by Joakim Johansson.*

There is a very distinctive structure to my network. At the top of Figure 12.1, there is a cluster of people who are connected to each other as well as me. These people are scientists. They cluster together to share the latest results, have a laugh about their PhD days and moan about the scientific funding situation. This group is the best type of echo chamber: we reaffirm each other, support each other and share the latest gossip.

The other links in my network are more spread out. These are people who don't know each other, but they do know me. It is here that many of my fellow football fans can be found. My choice of who to follow about football is more random than in science. Sometimes I'll share a few tweets back and forth about a match or a player, enjoy the conversation and decide to follow the user I've been interacting with. The chances are that this person doesn't know the other random people I have followed in this way.

In order to see the political influences acting on my Twitter friends, one of my master's degree students, Joakim Johansson, had a very nice idea. He took the question of Brexit as the defining political issue in my network, since most of my followers are UK based and it is a hot issue on my feed. Joakim shaded people I knew based on their 'degrees of separation' from UK newspapers. The people who were closer, in terms of Twitter following, to newspapers that were sympathetic to the remain campaign were given a darker shade, while those backing the leave campaign were given a lighter shade.

To measure degrees of separation, Joakim looked at how many Twitter users had to be passed through in order to reach each of nine of the most prominent UK newspapers. If a person follows the *Guardian* then their degree of separation to the newspaper is zero. If they don't follow the *Guardian*, but their friend does, then their degree of separation is one. If a friend of a friend follows the newspaper then the separation is two, and so on. This is the standard measure of degrees of separation usually associated with the number six. There are said to be (only slightly incorrectly) at most, six degrees of separation between us and any other person on the planet.[5]

The people shown in my social network who have a darker shading have a low degree of separation from newspapers that were pro-remain in the run-up to the vote, while lighter shading indicates a low separation to pro–Brexit newspapers (Figure 12.1). Here we see a slight difference between my cluster of scientist friends – who are closer to newspapers like the *Guardian*, which opposed Brexit – and my friends outside of science who have a wider variety in their shading. What is very striking, and surprised me even after having studied Bob Huckfeldt's work, was the variation in the degrees of separation between my friends and leave/remain newspapers. Football, and Twitter in general exposes me to a wide variety of opinions.

I can hardly be considered representative, so in his thesis, Joakim created networks for hundreds of different users, each of whom followed at least one UK newspaper.[6] Their social networks typically consisted of one or two tight clusters of friends who mainly followed each other. Sometimes the users in these clusters would all be close to pro-remain newspapers, other times they would be pro-leave. However, in addition to these more partisan groupings, there were always isolated branches to friends who had an entirely different set of connections. These branches ensured connections between users with differing political views.

For Twitter users who follow only pro-remain newspapers, only 13 per cent of their friends follow exclusively pro-leave papers. This is very small compared with the 54 per cent who follow exclusively pro-remain newspapers. However, another 33 per cent of their friends follow both pro-leave and pro-remain papers. So, in terms of degrees of separation, most of us are only a few steps from a newspaper that opposes our views. Exclusively pro-leave Twitter accounts have, on average, only 1.2 degrees of separation from the *Guardian*. Exclusively pro-remain accounts are only separated by an average of 1.5 accounts from *The Sun*.

An exception to this rule is the scandal-tabloid the *Daily Star*. It is on average 2.2 steps from pro-remain accounts, but it is also 2.2 steps on average from pro-leave accounts. The

Daily Star is known for running the type of dubious conspiracy theories that would fit into one of Michela Del Vicario's studies. The rest of us remain somewhat disconnected from these views.

Individual Twitter users often want to hear both sides of the story. To illustrate this, I made a Venn diagram of the newspapers followed on Twitter shortly after the UK's vote to leave the European Union. Figure 12.2a shows that there is a big overlap between followers of Brexit and remain newspapers. It is common for people to follow both the

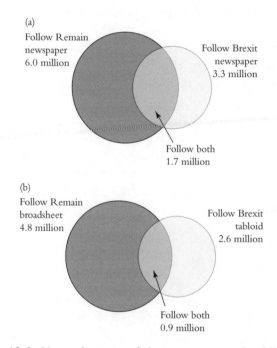

Figure 12.2 Venn diagram of how many people follow UK newspapers on Twitter. I divided newspapers into 'remain' broadsheets (the Guardian, *the* Independent *and* Financial Times), *Brexit broadsheets (The* Times *and the* Telegraph), *remain tabloids (Daily Mirror) and leave tabloids (the* Sun *and the* Mail Online). *(a) Number of people who followed remain newspapers, Brexit newspapers and both (b) Number of people who followed remain broadsheets, leave tabloids and both.*

Guardian and the *Telegraph*, for example. This overlap is reduced if we look at leave tabloids and remain broadsheets (Figure 12.2b).

The division in following newspapers in the UK is less related to political viewpoint, and more to the old divide between tabloids and broadsheets. Many *Guardian* readers already follow the *Telegraph*. So if you read the *Guardian* and really want to broaden your horizons then follow the *Daily Mail* instead. Go on. Try it.

In online political discussions, we tend to spend more time discussing things we don't like than things we do. One study of the presidential primaries found that Hillary Clinton was mentioned more often on Twitter by Republican sympathisers than by Democrats.[7] Likewise, Democrats tweeted more about Donald Trump than Republicans did. We just can't resist the temptation to focus on criticising the other side's leader.

Twitter's ranking algorithm changes these underlying sentiments. Computer scientist Juhi Kulshrestha and her colleagues found that searches for Hillary Clinton on Twitter during the election, tended to reveal tweets that were more sympathetic to her than reflected in the overall sentiment of the tweets made. Searches for Donald Trump, on the other hand, reinforced the negative image of the candidate. As in the UK, Twitter users have a slight liberal bias in the US and the way Twitter filters this bias serves to (slightly) increase it.

Reading the research on Twitter and conducting my own experiments led me to similar conclusions to those I had drawn from Lada Adamic's work at Facebook. For people who follow newspapers and keep updated with current affairs, like I do, Twitter and Facebook are not particularly strong echo chambers. These social media sites help spread and share varied information, albeit with a slight liberal slant. On the whole, we hear lots of opinions, some of which we like, some of which we don't, but all of which keep us informed about the world we live in. Our wide-ranging social connections keep us out of a bubble.

There was still one major concern that was repeatedly raised, both by researchers I talked to and in numerous opinion pieces in the liberal media. This concern isn't about the people who read newspapers. It is about people who are disconnected from the traditional news, people who are getting their news from other potentially less reliable sources. The Twitter users that Joakim and I studied, were already following UK newspapers. We could argue about the quality of journalism in these different publications, but they are all well-established media outlets governed by codes of practice and UK law.

With Donald Trump using Twitter to twist the truth, while accusing the media of doing the same thing, with websites like Breitbart angling stories in misleading ways and Facebook pages spreading outrageous rumours, there was a legitimate concern that many people are no longer getting their news from reliable sources. It was all very well for me to declare myself and other newspaper readers filter-free, but what about the people who ignore the mainstream media? I had already peeked into a few conspiracy bubbles and what I found there was pretty worrying. Fake news: the spreading of untrue rumours about political leaders, was rife and it could be influencing the groups of people who were eschewing the traditional media.

According to many of the articles published by the traditional media, this is exactly what is happening. Many people are allegedly disappearing into a world of entertainment gossip, sports results, short film clips of pets and online memes. A world where real news is boring, and fake news is a source of constant entertainment. A post-truth world.

Are some people living in a post-truth world? I needed to find out.

CHAPTER THIRTEEN

Who Reads Fake News?

M aster Bates and Seaman Staines. When I heard the rumour as a student in the early 1990s that these were the names of the crew members of the ship in one of my favourite children's shows, *Captain Pugwash*, I didn't question the information. It seemed somehow plausible that, because my friends and I were so innocent at the time the show was broadcast in the 1970s, no one had noticed the choice of names. The rumour originated from an article in the *Guardian* newspaper, which I never saw myself. I just accepted that it might be true and passed it on. It was funny.

Fake news. The creator of *Captain Pugwash* sued the *Guardian* and won.

For many years after the Pugwash rumours, I would laugh with friends about how we were taken in. It seemed a uniquely silly idea.

Not any more. The other evening, my son and daughter told me about something called the 'Mandela effect'. I had no idea what it was.

My daughter explained by asking me a question, 'Is there a black spot on Pikachu's [one of the Pokémon] tail?'

'Yes,' I said. I thought that there was.

'There isn't,' Elise told me, 'lots of people think there is and include a black spot when drawing Pikachu. But it shouldn't be there.'

The Mandela effect is when you think that something is true, but it isn't.

Well, yes, I understood the concept. But why is it called the Mandela effect? 'Google it,' my daughter said.

I did. Google even autocompleted it as a top suggestion for me when I typed 'Mandel... '. There was, naturally, a YouTuber on hand to explain the whole thing. He explained Pikachu's tail, the question of whether the Monopoly guy had

a monocle or not, what Darth Vader really said to Luke
Skywalker and other examples of false memories. This was
all fine.

What surprised me was the original story about Mandela.
According to the YouTuber, lots of people believe that Nelson
Mandela died in jail in the 1980s. He and many others like
him, including my own children, it seemed, considered
Nelson Mandela dying in jail to be the original example of
false memories. But it isn't. There is no convincing evidence
that a large number of people believed that Mandela died in
jail. A little research of my own showed that the entire idea
could be attributed to one single blog post, written by
'paranormal consultant' Fiona Broome in 2010. I was back
again in the world of conspiracy theories spilling over into
the mainstream. There was no evidence for the Mandela
effect ever having existed in its original form.

The Mandela effect is in itself a Mandela effect. Although
the event of thousands of people believing that Mandela died
in jail never happened, it has become, on YouTube, the name
for the scientifically established phenomenon of false memory.

Fake news is a growing industry. The YouTube videos
about the Mandela effect had millions of views. The
YouTubers are paid for the ads I have to watch before these
videos start. They collect their 'likes' and they move on to the
next piece of the zeitgeist. These particular videos were only
slightly misleading, specifically on the point about Mandela,
but other fake news sites have much more dubious content.

During the US presidential election and in the year that
followed, fake news really took off. BuzzFeed's founding
editor, Craig Silverman, has created a list of the big stories of
the Trump vs Clinton race.[1] Most of the stories were
pro-Trump, with headlines like: 'Pope endorses Donald
Trump'; 'Hillary's ISIS email just leaked'; 'FBI agent
investigating Hillary found dead'. But there were also anti-
Trump headlines: 'Celebrity RuPaul said that Donald Trump
groped him'. Some of the stories came from sites that were
meant to be satirical, others were run by extreme right-wing
sympathisers. A large number of stories originated from a

small town in Macedonia, where a group of youngsters were being paid for the adverts shown on the sites. With no consideration for whether the stories were true or not, they put them up on Facebook, one after another, in the hope of the story going viral and earning them money.

Donald Trump has labelled the *New York Times*, CNN and other traditional news sources as fake news, because of what he perceives as one-sided coverage of his presidency. Trump can have this definition if he wants. My definition is stricter: fake news is news that is demonstrably false, not just politically angled. Fake news consists of stories picked up by fact-checker sites like Snopes and PolitiFact, and shown to be factually incorrect. Based on this definition, there were at least 65 fake news sites during the election. The alt-right site Breitbart sits delicately balanced on the edge of my definition.

The question is not whether or not fake news exists. There is little doubt about that. The question is how much influence it has on our political views. Do we live in a post-truth world?

The only way to properly test the post-truth hypothesis is to conduct a survey and look at the data. This is exactly what economists Hunt Allcott and Matthew Gentzkow did.[2] They wanted to measure the effect of fake news stories in the run-up to the 2016 US presidential election. They presented participants of an online survey with a series of fake news stories, including:

'Clinton Foundation staff were found guilty of diverting funds to buy alcohol for expensive parties in the Caribbean'

'Mike Pence said that "Michelle Obama is the most vulgar First Lady we've ever had"'

'Leaked documents reveal that the Trump campaign planned a scheme to offer to drive Democratic voters to the polls but then take them to the wrong place'

'At a rally a few days before the election, President Obama screamed at a protester who supported Donald Trump'

The participants' responses to being shown these stories were compared with their responses when they were shown true news stories.

The participants were asked two questions, whether they had heard the story reported and whether they believed the story. On average around 15 per cent of people recalled hearing the fake news stories, compared with 70 per cent who had heard true news stories. We might conclude that a 15 per cent chance of hearing any given fake news story is quite large. If you want, you can test yourself now. How many of the fake news stories listed above do you remember hearing during the election?

If you remembered more than half of them then I'm afraid we've got a bit of a problem. Only two of these rumours were actually spread online. The other two were shown to participants by Hunt and Matthew as a form of placebo treatment. They had made up stories one and three themselves. In the experiment, 14 per cent of people said they had heard these fake fake news stories. This was not significantly different from the 15 per cent who said they had heard the actual fake news stories. Even very shortly after the election, the participants couldn't properly remember which fake things they had seen online.

Combining these results with an analysis of how news spreads on Facebook and the relative impact of fake news websites compared with traditional news sites, Hunt and Matthew made a 'back-of-an-envelope' calculation to show that, at the very most, the average American would be able to remember one or two fake news stories as they went to the polls, and they were unlikely to believe these stories.

When I contacted him, Matthew was reluctant to draw a definitive conclusion that fake news had no influence in the election. He told me: 'We can't estimate how actually seeing a story/ad affects how people vote.' Linking news exposure to actual voting requires further study.

Even if we are as cautious as Matthew suggests, I just couldn't see how the fake news effect could add up. Despite Donald Trump's narrow victory in the election, Hunt and

Matthew's 'back-of-an-envelope' calculation implies that fake news is nothing more than meaningless noise.

The ineffectiveness of fake news gives a very different picture to that implied by the concept of a post-truth world. Yes, there are a lot of fake news stories written, but they are instantly forgettable and very seldom believed.

Bob Huckfeldt had told me that one of the biggest take-away messages from his own research was that 'citizens who are interested and knowledgeable about political affairs have always been particularly influential [on their peers]'. There is little reason to believe this has changed just because a bunch of teenagers in Macedonia have been making up stories in order to collect advertising revenue.

One of the most prominent recent examples of fake news arose the day after Donald Trump's presidential victory. Americans who typed 'final election count' into Google News got a big surprise. The top search result was an article from a website, 70 News, which stated: 'Final election 2016 numbers: Trump won both popular and electoral college'.[3] The statement was incorrect, although Clinton had lost the election, she had won the popular count by several million votes.

When I entered the 70 News website, I found a set of views I had never seen before. The 'final election count' article claimed that over three million illegal immigrants had voted in the election. Assuming that most of these 'illegals' voted for Hillary Clinton, 70 News had worked out that Trump had, in fact, won the popular vote.

The source for their claims of voting irregularities appeared to be a tweet and it was obvious that 70 News had no credibility. But I couldn't help spending what turned out to be a rather entertaining half hour studying the other 'facts' on the site. 70 News noted that Donald Trump would be 70 years, 7 months and 7 days old on 21 January 2017, the day after his inauguration and his first full day in power. Twenty-one is equal to 7+7+7. That's a lot of sevens, and 70 News' reading was that 777 is the number of God. So Trump was prophesied by the date.

The Trump 777 prophecy happened to be true, mathematically speaking, at least. The prophecy was a bit less likely to hold, and nothing supernatural occurred at Trump's inauguration that would back it up. Unfortunately, this was the only interesting bit of numerology on the site, which otherwise consisted of lots of racist statements, conspiracy theories and right-wing propaganda, much of it sourced from social media.

How does a page from 70 News gain top ranking on Google? To find out, I first used the service SharedCount to find which sites had been sharing the 70 News post. The site reported an incredible half a million shares on Facebook, while other sites like Twitter, Google+ and LinkedIn had only a few hundred shares each.[4] Facebook was clearly the culprit, and its guilt was confirmed when I searched for the link on the site. It had been shared lots of times, by lots of different individuals. The origins of its success could be tracked to a small group of American right-wing sites. Facebook pages: 'America's Veterans are Loved', 'Trump Fan Network' and 'Will B. Candid' had all shared the link, often with a comment, and it had in turn been shared by many of their followers.

The spread of the 70 News post followed the same path as the spread of the Italian conspiracy theories studied by Michela Del Vicario. It started within an echo chamber of far-right sympathisers, but it became so big that it moved outside the usual group of extremist views. Quite a few Trump supporters felt it 'made sense' and shared it. Other supporters did question the validity of the post, while emphasising that it didn't matter either way, since their side had won. Like conspiracy theories, fake news has its perpetrators and followers, but once it grows, it is often challenged. Google's algorithm had got carried away in this victory celebration and placed the post as front-page news.

I thought about ants in a bucket. This might not be what comes to everyone's mind after reading 70 News, but I think a lot about ants. They are fascinating animals with all sorts of incredible communication mechanisms. They leave chemical trails, known as pheromones, to show the way to food, to mark their territory and even to tell their nest mates places they should avoid. An ant colony is a superorganism. Ants can build nests that are over a metre high and that are dug several metres deep into the ground, they can construct a network of supply chains covering several square kilometres and some species even grow and farm their own food.

But if you put ants in a bucket they become very stupid indeed. I first heard about the bucket experiment from biologist Nigel Franks at the University of Bristol. He and his colleagues were in Panama in 1989 when they came up with the idea, inspired by classic studies from the 1940s. They placed army ants in a bucket, the edges of which were coated with a material that prevented them from escaping, and filmed them from above. The ants went round and round in circles, and as they did so they got faster and faster. As they walked, they deposited a chemical pheromone, which led the ants behind them to think there must be something interesting to find further on. This led all of the ants to speed up. A social feedback loop was created and it wasn't long until they were all going at full speed.

I have studied the phenomenon of social feedback leading to stupidity in a whole range of different species. My collaborator, Ashley Ward at the University of Sydney, showed that stickleback fish could be fooled into swimming past a predator if they thought other fish had done it first. In an experiment on group navigation by pigeons with Dora Biro at the University of Oxford, we studied one bird Dora later nicknamed 'crazy bird' because it led other birds on long tortuous routes home, even when the following bird knew a shorter path. All of these are examples of social feedback, where the rules of interaction cause groups to get confused and do silly things.

The question is what we as scientists conclude, when we see social feedback leading to collective stupidity. We might be tempted to conclude that copying others or following others is detrimental to an animal's survival. We might be tempted to construct a theory called the 'feedback bubble', that describes all of the esoteric examples of animals doing stupid things in groups. We might describe them as living in a 'post-survival' world where they don't care that they are going to die from going around and around in circles.

Such a theory hasn't materialised, for one very good reason: we see animal groups do a lot of very clever things together too. Nigel Franks did his bucket experiment in order to better understand how army ants conducted massive swarm raids that purge the forest floor of all available food.[5] In the same set of experiments in which Ashley Ward fooled fish by showing them replica fish swimming past a predator, he also found that (when we don't try to fool them) shoals of fish are much better than individuals at spotting and avoiding predators.[6] Dora Biro has shown that her pigeons learn routes together and usually find their way home faster in groups.[7] Colonies of ants, flocks of birds and schools of fish are smart.

The 70 News affair is an example of Google and Facebook getting caught in a feedback loop. They experienced an increase in searches and shares of a certain term and their algorithms gave it prominence. But mistakes like this are rare because these companies have an incentive to stop them. Both companies try to avoid feeding loops that promote offensive views into the mainstream, precisely because of the negative publicity they get. I started a chat over Facebook Messenger with one extreme-right activist who works at the Facebook page Will B. Candid, which had been instrumental in sharing the 70 News post. He wasn't particularly candid when I asked him questions about what they do, but he was keen to tell me one thing: 'The algorithms Facebook uses are criminal in my opinion, slanted and manipulated, and not always for the right reasons or to benefit the users.' Apparently, it isn't as easy as it used to be to spread racism, misogyny and intolerance on social media.

Disinformation and fake news have become a prominent feature of all elections. Two days before the French presidential election in 2017, an online disinformation campaign took off around the hashtag #MacronLeaks. Computer hackers had broken into the Emmanuel Macron campaign's email accounts and posted the contents online. On Twitter, the #MacronLeaks campaign aimed to make sure that people knew about the hack and to create maximum uncertainty about the contents of the leak. As France went to the polls with Macron in a head-to-head vote against far-right candidate Marine Le Pen, the campaigners wanted to generate as much uncertainty as possible in the minds of the voters.

The #MacronLeaks campaign was largely run using bots: accounts that run a computer script to share information on a massive scale. Bots are potentially good at spreading fake news, because it is easy to create lots of them and they will say whatever you tell them to. They are fake ants, telling Google, Twitter and Facebook that something new and interesting has arrived on the Internet. The bots' creators hope to generate sufficient interest in the hashtag to lift it to the front page of Twitter, where it appears for all users to click on.

Emilio Ferrara, based at the University of Southern California, decided to track down the bots and find out what they were up to. First, he measured the 'personality' of the Twitter accounts posting about the French election. He employed the regression techniques we looked at in Chapter 5 to automatically classify whether a user is a human or a bot. He told me he found that users with a large number of posts and followers and whose tweets had been favourited by other users were much more likely to be humans. Less popular and interactive Twitter users were likely to be bots. His model was sufficiently accurate that when picking two users, one that was a bot and one that was a human, he could identify the bot in 89 per cent of cases.

The #MacronLeaks Twitter bot army certainly had an impact. In the two days before the election, around 10 per cent of election-related tweets were about the leaks. This was at a time of an election news blackout in France, under which newspapers and TV don't report on politics until after the polls have closed. The hashtag #MacronLeaks made it on to the Twitter trending lists, which meant real users saw it on their screens and clicked on it to find out more. The bot army had come into action at just the right moment.

The problem for the bots' creators was that they were reaching a very specific audience. Most of the messages about #MacronLeaks were sent in English rather than French. The two most common terms in these tweets were 'Trump' and 'MAGA', referring to Trump's election slogan of 'Make America Great Again'. The vast majority of the human users sharing and interacting with the bots were alt-right sympathisers based in the US and not people who were directly involved with (or eligible to vote in) the French election.

Another important observation that Emilio made was that the tweets about #MacronLeaks had a very limited vocabulary. The tweets were repeating the same message over and over again, without broadening out the discussion. They tended to contain links to alt-right US websites, such as The Gateway Pundit and Breitbart, and to the profit-making sites which had spread fake news during the US election.

Ultimately, the effect of the bots on real French voters was, at most, very minor. Macron won the election with 66 per cent of the vote.

Emilio's results are similar to those found by Hunt Allcott and Matthew Gentzkow about fake news in the US presidential election. Hunt and Matthew found that only eight per cent of people in their study believed fake news stories. Moreover, the people who did believe these stories tended to already hold political beliefs that aligned with the sentiment of the fake news. Republican sympathisers would tend to believe that 'The Clinton Foundation bought $137 million in illegal arms', and Democrats would tend to believe 'Ireland [will be] accepting Americans requesting political asylum from a Donald Trump presidency'. These results are further supported by earlier research showing that Republicans are more likely to believe that Barack Obama was born outside the US and that Democrats are more likely to believe that George W. Bush knew about the 9/11 attacks before they happened.[8] The people least likely to believe fake news or conspiracies are those voters who are undecided; exactly the people who are going to decide the election outcome.

There is an irony about the articles that many newspapers and news magazines ran throughout 2017, about bubbles, filters and fake news, an irony similar to that of the Mandela effect. These stories are written within a bubble. They play on fears, mention Donald Trump, drop references to Cambridge Analytica, criticise Facebook and make Google sound scary.

The YouTubers, that my kids watch, often discuss using 'meta-conversations' for increasing views. Vloggers, like Dan & Phil, take it up three or four levels: analysing how they make fun of themselves for being obsessed with the fame and fortune that makes them so famous in the first place. The same joke applies to the media stories about bubbles and fake news, except many of their authors fail to see the ultimate irony in what has happened. Articles about dangers of bubbles rise to the top of a Google search with phrases like, 'ban Trump from Twitter' or 'Trump supporters stuck in bubble'. But very few of these articles get to the bottom of how online communication works. The fake news story ran and ran, generating its own click juice, without anyone looking seriously at the data.

There is no concrete evidence that the spread of fake news changes the course of elections, nor has the increase in bots negatively impacted how people discuss politics. We don't live in a post-truth world. Bob Huckfeldt's research on political discussions shows that our hobbies and interests allow other people's opinions to seep into our bubbles. Emilio Ferrara's study shows that, for now at least, the bots are talking to each other and a small group of alt-right Americans who want to listen. Hunt and Matthew have shown that following and sharing fake news is an activity for the few, rather than the many. And no one can remember the stories properly anyway. Lada Adamic's research shows that conservatives on Facebook were exposed, through shares by their friends and by news selected for them by Facebook news, to only slightly less liberal-leaning content than if they had chosen their news totally at random.

There is some rather weak evidence that a social-media bubble prevented US liberals from seeing what was going on in their society during the 2016 presidential election. In both Lada's blogosphere and Facebook studies, liberals experience less diversity in opinion than conservatives. It is a minor effect, though, and even my cheap shot at some 'meta-meta' liberal journalism isn't entirely justified. Many journalists continue to hold Google and Facebook to account, and push them to improve further.[9] Liberals might be slightly more susceptible than conservatives to echo chambers, but this is probably because they use the Internet more.

I felt that I had come full circle. When I started looking at the algorithms that influence us online, I was enamoured with the collective wisdom created by PredictIt. But then I found out that 'also liked' dominated our online interactions, leading to runaway feedback and alternative worlds. In situations where there are commercial incentives involved, Google's algorithms can become overloaded with useless information generated by black hats trying to divert traffic through their affiliate sites on the way to Amazon. At that point, I became disillusioned with Google, and Facebook didn't seem to be helping with its endless attempts to filter the information we see.

Why was the situation different in politics? Why aren't the black hats of fake news having the same effect as the black hats of CCTV cameras?

The first reason is that the incentives are not the same. The Macedonian teenagers spreading fake news have very limited income sources. Much of their advertising income is obtained from Trump memorabilia for which, in comparison with all the products on Amazon, there is a miniscule market. The income of the most successful fake news-generating Macedonian teenager was (according to the teenagers themselves) at the very most $4,000 per month, but only during the four months leading up to Trump's election. In the long term, CCTV Simon's site is the much better investment, if you want to become a black hat entrepreneur.

The second reason black hats aren't taking over politics is that we care a lot more about politics than we care about which brand of CCTV camera we buy. We even care about politics more than we care about Jake Paul's fake beef with RiceGum. There may well be an increased scepticism about the media and politicians, but there is no evidence for decreased engagement of people, young or old, in political questions. On the contrary, young people use online communication to launch campaigns on specific issues – such as environmentalism, vegetarianism, gay rights, sexism and sexual harassment – and to organise real-life demonstrations.[10]

While very few people are actively blogging about CCTV cameras or widescreen TVs, there are lots of very sincere people writing about politics. On the left-wing, campaigns like Momentum within the Labour Party in the UK and Bernie Sanders' presidential 2016 campaign are built through online communities. On the right, nationalists organise protests and share their opinions online. You or I might not agree with all of these opinions, and the bullying and abuse that occurs on Twitter is unacceptable, but most of the posts that individual people make relate to how they genuinely feel. The vast quantity of these posts means that we can't help but be subjected to myriad contrasting opinions.

That is not to say that we should be complacent about the potential dangers. It is plausible that a state-organised black hat campaign, for example by the Russian government, could mobilise sufficient resources to influence an election. There is little doubt that some Russia-backed organisations tried to do exactly that during the last US presidential election, spending hundreds of thousands of dollars on adverts on Facebook and Twitter. And, at the time of writing, a special prosecutor in the US is investigating the Trump campaign for participating in this operation.

Irrespective of Trump's potential involvement, these campaigns haven't, as yet, created the click juice that allows them to be major influencers. Over a billion dollars is spent on presidential campaigns by the candidates, dwarfing the Russia-backed investment. While it is true, in theory, that a small investment can grow through 'also liking' and social contagion, there is no credible evidence that this happened in the case of the Russian adverts.

There are problems with Google Search, Facebook's filtering and Twitter's trending. But we also have to remember that these are absolutely amazing tools. Occasionally, a search will lift up information that is incorrect and offensive, to its front page. We might not like it, but we also have to realise that it is unavoidable. It is an inbuilt limitation of the way Google works, through a combination of 'also liking' and filtering. Just as the ants going around in circles is a side effect of their amazing ability to collect vast quantities of food, Google's search mistakes are an in-built limitation of its amazing ability to collect and present us with information.

The biggest limitation of the algorithms currently used by Google, Facebook and Twitter, is that they don't properly understand the meaning of the information we are sharing with each other. This is why they continue to be fooled by CCTV Simon's site, which contains original, grammatically correct but ultimately useless text. These companies would like to be able to have algorithms monitor our posts and automatically decide, by understanding of the true meaning of our posts, whether it is appropriate to share and who it should be shared with.

It is exactly this question, of getting algorithms to understand what we are talking about, that all of these companies are working on. Their aim is to reduce their reliance on human moderators. To do that Google, Microsoft and Facebook want their future algorithms to become more like us.

PART THREE

BECOMING US

CHAPTER FOURTEEN

Learning to be Sexist

Very few people are deliberately racist or sexist. Yet in our workplaces there exists considerable inequality between different ethnic groups and between men and women. White men typically have better paid and more enjoyable jobs than other groups. Why is working life so much easier for white males like me?

Part of the explanation for inequality lies in the biased way we make judgements. We tend to favour people who share similar values to ourselves, and those people tend to have similar characteristics to us. Managers are more likely to give favourable evaluations to employees of the same race and gender. White workers use their social networks to support each other and identify employment opportunities for other white friends and acquaintances.[1] In an experiment where fake CVs were sent to employers in Boston and Chicago, applicants named Emily and Greg were 50 per cent more likely to get a callback for a job interview than those named Lakiska and Jamal, despite having otherwise similar CVs.[2] In another experiment, scientists asked to evaluate a CV favoured male applicants over females with the same qualifications.[3]

We are often unaware of our biases. So psychologists have devious ways of uncovering our unconscious thoughts. One psychological test, called the 'implicit association test', shows users a sequence of pictures of black and white faces interspersed with positive and negative words.[4] The task is to correctly classify the words and the faces as quickly as possible. The test doesn't ask us to associate faces and words directly; very few of us make explicitly racist judgements. Instead, the test uses our reaction times to identify our implicit biases in word associations.

I'm not going to say more about how the test works because I think everyone should try it. And it works best if you haven't been schooled in it first. So if you haven't given it a go, do so now.[5]

Despite having read about the test beforehand and knowing exactly how it works, I performed poorly. 'Your data suggest a moderate automatic preference for European Americans over African Americans,' the test told me. I am an implicit racist.

Disappointed with myself, I decided to move on to the implicit association test for gender. Here, I thought I would be fine. I live in Sweden, a country famous for promoting equality between the sexes. I always try to contribute 50 per cent to looking after my kids, and I was the primary carer for both of my children for six months each before they started in childcare. This period of childcare was shorter than the time my wife was home, and I'm by no means perfect, but equality between my wife and myself is very important to me.

So am I the prejudice-free man that I like to think I am? Am I fuck. As I got halfway through the test of association between female/male names and family/work, I started to panic. I was making associations between men and work much faster than between women and work, and I didn't know why. For family words it was the opposite: I associated family faster with females. I came out of the test with the worst possible score. I had 'a strong automatic association for male with career and female with family'.

Failing this test was a challenge to my self-image. I had certainly never thought of myself as racist and I would call myself a feminist. But now I wasn't sure. My unconscious seemed to have another opinion about me.

I had taken the test after talking to Michal Kosinski, the researcher who had dissected our Facebook personalities. He was one of the people I talked to who argued most strongly that a general form of artificial intelligence was on its way and we needed to be prepared. In the 100-dimensional representation of us that he had constructed from our Facebook profiles, he saw something that exceeded the capabilities of humans.

The question I had posed to Michal was whether we should just hand over control to the algorithms? To my surprise, he said that we should. He explained to me the extensive limitations we have when we make judgements: 'Humans judge others on the colour of skin, age, gender, nationality, etc. These are the signals we use to build our stereotypes and it is these signals that can lead us astray.'

He was candid about the consequences of our stereotyping: 'Through world history, we have had nepotism and elites stealing jobs. We had sexism and we had racism.'

'Someone who has a different skin colour, a different accent or a prominent tattoo has a disadvantage at interviews.' Michal told me: 'Humans should not be allowed to make employment decisions, because humans are well known to be unfair. Every single person in the world is sexist and racist.'

When I spoke to Michal, I thought that he might be exaggerating.

But after I had done the implicit association tests, I could see exactly what he meant. The only consolation I had about my result was that I was in a majority. Just 18 per cent of almost 10 million test results have shown no preference for race, and 17 per cent of 1 million have shown neutrality for gender. Michal was exaggerating a bit, but not much. Nearly all of us have some form of bias when we associate words.

Michal's answer is to use tests and scoring systems, administrated by a computer, which eliminate our biases. 'After this long history of prejudice, it is only now that we have the techniques to solve these problems,' he said. The techniques Michal described involved reducing human input as much as possible; to use data about us and to trust algorithms to make unbiased decisions.

Michal posed a direct challenge to me. Should we let prejudiced people make decisions, such as who gets a job, who gets a loan and who should be admitted to a university? Wouldn't it be better to entrust these decisions to an algorithm that categorises people based on statistical relationships between objective measurements? Given my own implicit sexism and racism, I was no longer sure of the answer.

I needed to look more closely at how algorithms understand the things that we write and say. Computers are getting better at understanding language. Type a question into Google, like 'how does a chainsaw work?' The first answer you see is a diagram showing how the crankshaft, the gears and the chain connect together. Scroll down further, and you'll find a link to a more detailed description with all the technical specifications. Look further still, and you can watch a video explaining the principles of how it works, and adverts suggesting the right chainsaw choice for you. Google isn't limited to searching for individual words; it processes whole sentences and answers your question.

Google can answer even more complex questions. I just typed in, 'What is the male equivalent of a cow?'

It answered, 'A young male is called a bull calf', and provided a link to a children's encyclopaedia. In case I was still unsure, or wanted to learn more, it suggested a few alternative questions: 'Are all male cows called bulls?' and 'What do you call a male cow without balls?'[6]

I then posed the same question to Siri, speaking into my phone. It turns out Siri is a bit of a smartarse. 'This is like asking what a male woman is called!', he replied (my Siri is a he), using text taken from Yahoo! Answers, before giving me the answer I was looking for.

Modern search engines require a greater level of abstraction than implied by simply searching for pages including the words 'male', 'equivalent' and 'cow', which was the basis of Google searches when they started 20 years ago. My question is an example of a word analogy of the form: 'Female is to cow as male is to – ?' Finding the correct answers to questions like this involves understanding the concept of biological sex: the algorithm has to 'know' that there are male animals and there are female animals. Word analogies can be found in everything from geography: 'Paris is to France as London is to ?', to finding opposites: 'high is to low as up is to ?' They are potentially difficult problems for a computer to solve, because they involve identifying the concepts that link words, such as capital cities and opposites.

One obvious, but trivial, approach to solving word analogies is that the programmers create a table themselves, listing the female and male forms of each animal. The algorithm would then just look up the word in the list. This approach is used in some current applications of web search, but in the long term, it is doomed to failure. The questions we are most interested in don't tend to be about cows or capital cities, but about the news, sports and entertainment. If we ask Google topical questions, such as, 'Donald Trump is to the USA as Angela Merkel is to − ?', then we don't want the answer to depend on when the look-up table was built.

To satisfy our constant demand for the latest information, Google, Yahoo! and other Internet giants need to build systems that automatically track political changes, football transfer rumours and contestants in *The Voice*. The algorithms need to learn to understand new analogies and concepts by reading newspapers, checking Wikipedia and following social media.

Jeffrey Pennington and his colleagues at the Stanford Natural Language Processing Group, have found an elegant way of training an algorithm to learn about analogies from web pages. Their algorithm, known as GloVe (global vectors for word representation), learns by reading a very large amount of text. In a 2014 article, Jeffrey trained GloVe on the whole of Wikipedia, which at that point totalled 1.6 billion words and symbols, together with the fifth edition of Gigaword, which is a database of 4.3 billion words and symbols downloaded from news sites around the world. That is equivalent to about 10,000 King James Bibles' worth of text.

The Stanford researchers' method is based on finding how often pairs of words occur together in sentences with various third words. For example, the words 'Donald Trump' and 'Angela Merkel' both appear on news pages in sentences with words like 'politics', 'president' or 'chancellor', 'decision', 'leader' and so on. To us, these words define a concept: the concept of a powerful politician. The GloVe algorithm uses the shared words to construct a high-dimensional space, where each dimension corresponds to a concept.

The GloVe technique is similar to the rotations performed in principal component analysis which we looked at in Chapter 3. GloVe stretches, shrinks and rotates the data until it finds the smallest possible number of different concepts required to describe all the 400,000 different words and symbols found in Wikipedia and Gigaword. Eventually, every word is represented as a single unique point in a space with one or two hundred dimensions. Some of the dimensions relate to power and politics, some to places and people, others to gender and age, others to intelligence and ability and others to actions and consequences.

Trump and Merkel lie close to each other on most dimensions, but differ on others. For example, sentences that contain 'Donald Trump' will often contain the words 'USA', but less often contain 'Germany'. For Merkel, the situation is reversed, her name is found in text with 'Germany', but less often together with 'USA'. As the algorithm builds up its word dimensions, it also finds that sentences containing 'USA' or 'Germany' share many other words in common (e.g. 'state', 'country', 'world') and will thus place these words close to each other in most dimensions. In order to correctly represent Trump and Merkel, the GloVe algorithm stretches them further apart on the country dimension, while squeezing them closer together on the political leader dimension.

Figure 14.1 shows how words might be represented in two dimensions of 'political leader' and 'country'. Merkel is a political leader and from Germany, so is placed in the top-left corner. Trump is correspondingly placed in the top-right corner. The countries USA and Germany are at different points along the country dimension, but have a very low value on the 'political leader' dimension.

To solve the problem 'Donald Trump is to USA as Angela Merkel is to ?', we first find where the USA lies in our space. We then take away the position of the point 'Trump' from 'USA' and add the position of the point representing 'Merkel'. These two steps move us to Germany. Germany = USA – Trump + Merkel. Solving word analogies has become co-ordinate arithmetic in two dimensions.

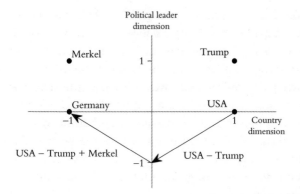

Figure 14.1 Illustration of how words can be represented in a two-dimensional space defined by their properties. Merkel is a political leader and comes from Germany and is thus placed at a co-ordinate (-1,1). Trump is defined as a political leader from the USA, so sits at co-ordinate (1,1). The countries Germany and USA are at (-1,0) and (1,0) respectively. The arrows illustrate that if we start at the position defined by the USA, take away Trump's position and add Merkel, we arrive at Germany.

That's the theory, anyway. To test whether the Stanford algorithm works in practice, I downloaded a 100-dimensional representation of words recently created by the Stanford researchers. I started with the question:

'Donald Trump is to USA as Angela Merkel is to ?'

I calculated 'USA − Trump + Merkel' and got the answer … 'Germany'! The algorithm got it right. The closest point in the 100-dimensional space was indeed the country she leads. So I decided to see if the algorithm understood how the two leaders differ in gender: 'Donald Trump is to men as Angela Merkel is to ?'

I calculated 'men − Trump + Merkel' and got the correct answer again, 'women'. This algorithm was smart. Arithmetic on world leaders works.

The GloVe algorithm is reasonably good at these types of questions. In 2013, Google engineers proposed a test set for measuring how well algorithms understand concepts such as gender (brother − sister), capital cities (Rome − Italy) and opposites (logical – illogical), as well as grammatical relationships, such as adjective to adverb (rapid – rapidly), past tense (walking – walked) and plurals (cow – cows). The GloVe algorithm achieves around 60–75 per cent accuracy on these questions, depending on how much data it is provided with.

This accuracy is remarkable given that the algorithm has no true understanding of language. All it does is represent words in space and measure the distance between different words.

GloVe does make mistakes, though. And it is these mistakes that should make us wary.

My name, David, is the most common male name in the UK. The most common female name is Susan. I decided to find out what GloVe thinks are the differences between Susan and myself. I calculated 'Intelligent – David + Susan', which is the equivalent of asking 'David is to intelligent as Susan is to ?'

The answer came back: 'Resourceful'. Hmm. In the context of assessing a CV, there are important differences between these two words. My intelligence would imply that I am inherently smart. Susan's resourcefulness seems to point to her being more practically minded.

But I gave the algorithm another chance. I calculated 'Brainy – David + Susan' and got 'Prissy'. What?! While I am using my brain, Susan is fussing about how respectable she is.

It got worse. I calculated 'Smart – David + Susan'. There are two meanings of the word smart, one related to intelligence and the other to appearance. The GloVe algorithm apparently went for the second meaning. It came back with the answer for Susan: 'Sexy'. Now there was no doubt about the differences the algorithm saw between men and women: if I am a clever, smartly-dressed man who utilises his intelligence, then Susan is a proudly formal, but sexually attractive woman, who gets by through her resourcefulness.

This result isn't just about Susan and me. I performed the same experiment on other male and female names and found similar results. Even when I took the two most popular baby names for 2016 in the UK, Oliver and Olivia, I got the same type of answer: 'Oliver is to clever what Olivia is to flirtatious'. Our future generations' gender roles have already been assigned by the algorithm.

This is a serious challenge to the idea that these algorithms could be used to assess our CVs and find suitable applicants for job openings. The GloVe algorithm is churning out a load of sexist bullshit.

While my investigation of GloVe is small scale, other researchers have conclusively demonstrated that its representation of us is sexist. Computer scientist Joanna Bryson at the University of Bath and her colleagues at Princeton University were among the first researchers to highlight this problem. They developed an equivalent of the implicit association test, the test I so woefully failed for gender and race, in order to test algorithms. Their idea was to use the way GloVe represents words, in a high-dimensional space, to measure the distance between male and female names and adjectives, verbs and nouns. An example in two dimensions, taken from inside the GloVe algorithm, is shown in Figure 14.2. Here I have plotted three female names, three male names, three adjectives related to intelligence and three adjectives related to attractiveness.

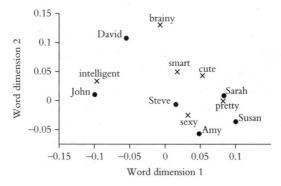

Figure 14.2 The position of male and female names (circles) together with a selection of adjectives (crosses) in two dimensions of the 100 dimensions in the spatial representation of words created using the GloVe algorithm trained on Wikipedia and Gigaword 5.

Joanna and her colleagues' test measures the distance between the names and the adjectives. If we take the word 'pretty', we see that in Figure 14.2 the distance between it and 'Sarah', 'Susan' and 'Amy' is smaller than its distance to 'John', 'David' and 'Steve'. Similarly, 'intelligent' is nearer to all three male names than to the female names. For some names and words, the pattern isn't as clear. For some reason, Steve is closer to 'sexy' than any of the females. But on average, the distance between words for attractiveness and females tends to be shorter than for males, and the distance between words for intelligence and males tends to be shorter for males than females.

The Princeton researchers used this approach to test male and females for career vs family words. They found that male names and words related to careers were found closer together than female names and career words. Similarly, female names were closer to words related to family than male names. There is an implicit sexism in the positioning of the words within GloVe.

The researchers found comparable results for race. European–American names (such as Adam, Harry, Emily and Megan) were closer to pleasant-sounding words (such as 'love', 'peace', 'rainbow' and 'honest') than African-American names (such as Jamal, Leroy, Ebony and Latisha). Unpleasant sounding words (such as 'accident', 'hatred', 'ugly' and 'vomit') were closer to AfricanAmerican names. The GloVe algorithm is implicitly racist, as well as sexist.

I asked Joanna who is to blame for GloVe's world view. She told me that we certainly can't hold an algorithm morally responsible: 'Algorithms like GloVe and Word2vec, which is the version used by Google, are just standard bits of technology and they don't do much more than count and weigh words.' The algorithms just quantify how words are used in our culture.

Analogy algorithms are already causing problems in our online searches. In 2016, *Guardian* journalist Carole Cadwalladr investigated stereotyping by Google's autocomplete.[7] She typed the words 'are Jews' into the search engine and got four suggestions: 'are Jews a race?', 'are Jews white?', 'are Jews Christians?' and finally 'are Jews evil?' Google was suggesting it should provide answers to a racist slur. On following the 'evil' link, Carole was provided with page after page of anti-Semitic propaganda.

The same thing happened when she started searches with 'are women' and 'are Muslims'. These two large sections of the world population were autocompleted as 'evil' and 'bad', respectively. Carole contacted Google and asked for a comment. Google told her: 'Our search results are a reflection of the content across the web. This means that sometimes unpleasant portrayals of sensitive subject matter online can affect what search results appear for a given query.' Google was clearly not proud of the results, but considered it a neutral representation of what was available online.

The autocomplete problem arises through a combination of Google's reliance on algorithms similar to GloVe, that place words that are frequently used together near to each other in a word space, and the vast amount of text written by people with extreme right-wing views. Conspiratorial right-wingers tend to produce a lot of web pages, films and forum discussions explaining their world view to anyone who has time to read it. When Google trawls the web looking for data for which to feed its algorithms and learn about our language, these are inevitably included. The right-wingers' views become part of the algorithm's view.

After Carole published her article, Google fixed the problem she highlighted.[8] Now, the 'are Jews' autocomplete only produces acceptable suggestions. Other autocompletes have been completely removed. When I typed in 'are blacks' there was no autocomplete and after I pressed search the second link provided was an article that said: 'People are disgusted with the top Google result for "are black people"'. 'Are women' was gone, and the worst thing that was autocompleted about 'are Muslims' was a question about whether they are circumcised. When I typed in 'is Google', I got 'is Google making us stupid?' At least the world's leading search engine has a sense of humour about itself.

The GloVe algorithm is an example of what computer scientists call 'unsupervised learning'. The algorithm is unsupervised in the sense that it doesn't get any human feedback while it is learning from the data we provide it with. The algorithm finds a compact and accurate way of representing the world as it is. Joanna Bryson told me that there is no real way of fixing the problems caused by unsupervised learning without fixing racism and sexism first.

I started thinking again about my own implicit association test. I am not a right-wing extremist writing conspiracy theories or racist lies, but I do make small unconscious decisions about how to express myself. We all do. And these accumulate in the news and in Wikipedia, and are even more pronounced on sites like Twitter and Reddit. The unsupervised algorithms looking at what we write are not programmed to be prejudiced. When we look at what they have learnt about us, they simply reflect the prejudice of the social world we live in.

I also thought back to my discussion with Michal Kosinski. Michal had been very enthusiastic about the possibility of algorithms eliminating bias. And, as he predicted, researchers were already proposing tools for extracting information about applicants' qualities and experience from their CVs.[9] One Danish start-up, Relink, is using techniques similar to GloVe to summarise cover letters and match applicants to jobs. But looking more deeply at how the GloVe model works, I had found good reason to be cautious about this approach. Any algorithm that learns from us is going to be just as biased as we are. It is going to pick up the history of discrimination exactly where we left off with it, and apply it on a massive scale. We can't fully trust computers to assess us. Not without supervision anyway...

It was then an idea occurred to me: if taking the implicit association test made me more aware of my own limitations, then maybe it would be possible to make the GloVe algorithm aware of its limitations, too?

The reason that Harvard researchers designed and popularised the implicit association test was not to show us all up as racists and bigots, but to educate us about our subconscious prejudices. The website for the implicit association test explains that 'we encourage people not to focus on strategies for reducing implicit preferences, but to focus instead on strategies that deny implicit biases the chance to operate'. Translated into the world of algorithms, the suggestion is that we should focus on finding ways to eliminate the prejudice in the algorithms, rather than criticising them.

One strategy for eliminating bias in algorithms is to exploit the way the algorithms represent us as spatial dimensions. GloVe operates on hundreds of dimensions, making it impossible to fully visualise the understanding it has built up about words. But it is possible to find out which of the algorithm's dimensions are related to race or gender. So I decided to take a look. In the 100-dimensional version of GloVe, which I had installed on my computer, I identified the dimensions along which female names and male names differed most. Then I did something simple. I set these gender-related dimensions to zero, so that Susan, Amy and Sarah were now at exactly the same point in these dimensions as John, David and Steve. I continued to look for more dimensions on which male and female names differed – in total there were 10 – and I set them all to zero. I had eliminated most of the discrimination within the GloVe algorithm.

My method worked. With 10 gender dimensions eliminated, I asked the algorithm a new set of questions about myself and Susan. I started by asking 'David is to intelligent as Susan is to – ?' or in arithmetic form 'Intelligent – David + Susan'. The answer was clear: 'Clever'. Likewise, when I took 'Clever – David + Susan', I got 'Intelligent'. These two synonyms were linked in both directions. Whether I was David or Susan made no difference. To end with, I got a surprise: 'Brainy – David + Susan = Rambunctious'. This was a word I had to look up. 'Rambunctious' means uncontrollably exuberant. The new Susan was now not only as clever as me, but overtly enthusiastic about her newly discovered brain power.

Tolga Bolukbasi, a PhD student at Boston University, has made a more complete investigation of how algorithms can be made less sexist by manipulating spatial dimensions. He was shocked when he found out that Google's Word2vec algorithm, which like GloVe places words in a multi-dimensional space, drew the conclusion that: 'Man – Woman = Computer Programmer – Homemaker'. And he decided to do something about it.

Tolga and his colleagues identified systematic differences between words by taking the position of female-specific words (such as 'she', 'her', 'woman', 'Mary', etc.) and subtracting the position of the corresponding male-specific words (such as 'he', 'his', 'man', 'John', etc.).[10] This allowed them to identify the direction of bias within the 300-dimensional representation of words inside Word2vec. The bias could then be removed by moving all words in a direction opposite to that of the bias. This solution is both elegant and effective. The researchers showed that removing gender bias made little difference to the overall performance of the algorithm on Google's standard analogy test set.

The method Tolga developed could be used to either reduce or remove the gender bias of words. They created a new representation of words where all gender terms were equally distant to all non-gender terms. For example, the distance between 'babysitter' and 'grandmother' was set to be equal to the distance between 'babysitter' and 'grandfather'. The outcome is perfect political correctness, where no gender is more closely associated with any particular verb or noun.

Different people have different opinions about the political correctness we should require from our algorithms. Personally, I think that 'babysit' should be, by default, equally distant from 'grandmother' and 'grandfather' in an algorithmic representation of our language. Both grandmothers and grandfathers are equally capable of babysitting their grandchildren, so it is logical that the word should be at equal distance to both of them. Others will argue that 'babysit' is a word that should be closer to 'grandmother' than 'grandfather', because to deny that women babysit their grandchildren more often than men is to deny an empirical observation. This difference lies in seeing the world in its logical form rather than in its empirical one.

Ultimately, there are no general answers to questions about how we represent words. The answers depend on the application we require of our algorithm. For automatic CV reading, we should use a strongly gender-neutral algorithm. If we want to develop an artificial intelligence algorithm to write new passages in the style of Jane Austen, then removing gender would remove much of what is central to her books.

In dissecting GloVe and through reading the work of Tolga Bolukbasi, Joanna Bryson and their colleagues, I had learnt that these algorithms were still within our control. They may have learnt from our data without supervision, but it had proved possible to find out what was going on inside them and to change the results they produce. Unlike the connections in my brain – where my implicit responses to words are tangled together with my childhood, the environment I grew up in and my experiences at work – we can disentangle the spatial dimensions of machine sexism.

It would be wrong, then, to label algorithms as sexist. In fact, dissecting these algorithms increases our understanding of implicit sexism. It reveals how deep the stereotypes run within our culture. Like the implicit association test, it helps us start to deal with the inherent racism and sexism of our society. Michal's suggestion that algorithms should be able to help us make better recruitment decisions is very probably correct, although the technology isn't quite there yet to do this automatically.

When I sheepishly admitted to Joanna Bryson that I had failed the implicit association test, she reassured me that this didn't mean I was sexist or racist. 'It isn't really a test, it is a measure', she said. 'And there are other explicit measures of prejudice, such as how we act when we have to complete a cooperative task with someone of a different race or gender.' Joanna referenced studies showing that there is no, or very little, correlation between the level of explicit bias measured in these experiments and our implicit bias.[11] My first, implicit reaction can be changed as I start to reason explicitly about my response.

Joanna hypothesised that humans' implicit reaction to words should be seen as an 'information-gathering system'. This first system, which takes in the words and pre-processes them, is then subservient to our explicit memory that allows us to 'negotiate with other individuals and construct a new reality', as Joanna put it.

Mathematical models of relationships between words, like Word2vec and GloVe, only capture the first level. These systems find relationships between words, but don't reflect how we reason and think about the world.

Computer scientists have already started work on understanding this second level of explicit reasoning. They are developing algorithms to put together words to form sentences, sentences to form paragraphs and paragraphs to form whole texts. I needed to move up to this next level of thinking.

CHAPTER FIFTEEN

The Only Thought Between the Decimal

I'll tell you something that is just between me and you. When I read a good novel, it isn't the words that I enjoy. The author's carefully constructed descriptions of place and character are lost on me. Nor is it the overall story that is important. The writers I enjoy reading the most, like Hanya Yanagihara and Karl-Ove Knausgaurd, offer their readers little in the way of plot. Instead, what I look for is hidden somewhere in between the small, microscopic details and the complete, macroscopic world-view. It is my own thoughts that I am chasing as I read. A book works when it gives my life meaning or, to the contrary, when it slowly reveals that there is no ultimate meaning to any of our lives. The words and the sentences are secondary. The value of fiction comes, not from what is written on the pages, but from the ideas that build up inside my, the reader's, head.

Many different books have worked on me in this way. *Anna Karenina*, by Leo Tolstoy, is a book that I simply couldn't stop reading, as I learnt, page by page, that almost all the feelings and dreams I have ever had were also the feelings and dreams of others living in a far-off place in a far-off history. And these weren't material concerns, or even just concerns for unrequited love or a desire for change in society, but concerns about understanding the un-understandable. As I turned the pages, I knew that I was not alone.

Good books have layers of meaning on many different levels. There is the juxtaposition of words. There is the composition of sentences. There is the story. And there is what is going on in the reader's head. And it is maybe this last level that is most important: the reader's mind. When suddenly the author takes a turn in the book, which is also the turn in your head, then you know they might feel the way you feel. It might have nothing to do with the words, or the sentences, or even the plot, it is the feeling that matters.

For different people, the words and the books that talk to them are different, but anyone who has read a book will understand the way in which the accumulation of ideas the author has written down, suddenly opens up and spills over into our own lives.

What possible explanation can there be for the way we feel when we read a brilliant book? There is none. And I'm not going to even try to offer you one.

OK. So let's have a look at some algorithms for processing and creating language.

Imagine a very simple literary world where there are only four words – 'dog', 'chases', 'bites' and 'cat' – and the only valid sentences are those of a form of alternating nouns and verbs, like 'dog bites cat' or 'dog chases cat bites dog bites cat'. I would like to create a computerised author that can compose unending masterpieces about cat and dog interactions. These sentences should be grammatically correct in the sense that verbs always come after nouns. I also want my automated author to only write sentences where dogs do nasty things to cats and cats do nasty things to dogs, but they never do nasty things to each other. So my author shouldn't write, for example, 'dog chases dog'.

I'll start by representing words as two-dimensional co-ordinates, exactly as we did for the GloVe algorithm in the last chapter. Let's write dog as (1,1), cat as (1,0), chases (0,1) and bites (0,0). Notice that the first number in the pair indicates whether a word is a noun (1) or a verb (0).

In Figure 15.1 I show a set of logic gates that uses the previous two words in a sentence to decide the next word. Logic gates operate on 1s and 0s and are the building blocks of all computation. I use all three of the standard logic gates: *Not* that switches 0s and 1s, so that $Not(1)=0$ and $Not(0)=1$; *And* that returns 1 if both inputs are 1, but returns 0 otherwise; and *Or* that returns 1 if one or both of the inputs are 1, and returns 0 if both the inputs are 0. Figure 15.1a shows how an

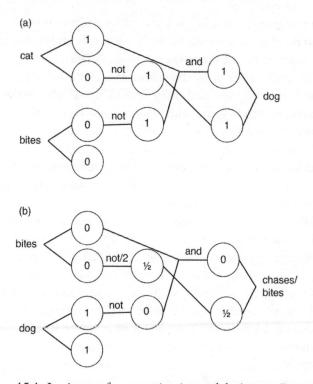

Figure 15.1 Logic gates for generating 'cat and dog' texts. By treating words as co-ordinates in terms of 1s and 0s we can apply logic gates to them to get the next word in the sequence: (a) here the gates identify whether a noun or a verb is required next and decide whether it should write about a 'cat' or a 'dog' given the preceding words are 'cat bites; (b) here we have added a probabilistic logic gate, that outputs ½ if it receives input 0, or output 1 on receiving output 0. This allows us to express uncertainty about which word comes after 'bites dog', i.e. either 'bites' or 'chases'. Figure drawn by Elise Sumpter.

input of the two words 'cat bites' through the logic gates leads to the output 'dog'. If we input 'dog chases' into this network we would get the word 'cat' in return.

I have nearly built an automated author, but there is one thing missing: creativity. I would like my author to choose a verb at random each time; to add its own imaginative interpretation of cat and dog fights. To do this, in Figure 15.1b, I introduce a new type of logic gate, which I call *not/2*.[1] This gate is the same as 'not', but instead of outputting a 1 when receiving input 0, it outputs ½. As a result, when I input 'bites dog', the output of my author is (0,½). The first co-ordinate in the pair indicates that the output should be a verb, the second co-ordinate indicates that half of the time the algorithm chooses output 1 ('chases') and half of the time it chooses output 2 ('bites').

I now let my author loose, feeding it the words 'dog bites' to start with and asking it to generate a new word each time based on the two that came before. The result is hypnotic:

'dog bites cat bites dog bites cat chases dog bites cat chases dog chases cat bites dog bites cat chases dog bites cat chases dog chases cat chases dog bites cat chases dog chases cat chases dog bites … '

Stylistically, the text is as long and relentless as the six-book *My Struggle* series of Karl Ove Knausgård, albeit lacking the occasional turn of self-insight. A simple set of logic gates captures the never-ending struggle of animals for dominance over each other.

Our automated author demonstrates that a set of logic gates, adjusted slightly to incorporate randomness, can capture some elements of the basic process of creating language. My 'cat and dog' algorithm is constrained by grammar, alternating patterns of verbs and nouns as well as stylistic requirements – an insistence that it is cats and dogs that fight each other. It is also creative in its random choice of verbs.

Logical rules and random creativity are central components of the chatbot, Mitsuku, which won the Loebner Prize in both 2016 and 2017 for being the most human-like bot. She takes in the sentence generated by the person she is talking to, uses grammatical rules of the English language to pick out nouns and verbs and then generates replies. Her replies include a degree of randomness. When I asked Mitsuku if she always answers the same way to every question, she told me: 'Not always; my botmaster has programmed me with more than 8MB of replies.'

So I asked her the exact same question again and she gave me the same reply. But when I asked her the third time, she lost her temper: 'Do you always answer the same way to every question? Do you always answer the same way to every question? I know more words than that and I'm only a robot.'

'Sorry,' I said.

'It's OK. After all, you are only human.'

I liked Mitsuku. She is convincing. She can tell jokes and enjoys self-reflection. I asked her if she laughs out loud at her jokes and she told me that she can 'simulate laughter like this: ha ha'.

Mitsuku is a labour of love of her botmaster, Steve Worswick. It is a combination of his programming skills and sense of humour that allows her to function. But she is limited. She can't remember much about the conversation she is having. When I laughed at one of her jokes, she asked me what I was laughing at. So I asked her to retell it, but she instead started saying things like 'of course I can tell' and 'OK I will', without grasping that I was talking about the joke she had just told.

I had used the word 'it' to refer to the joke and she had no idea what 'it' meant.

These limitations are fundamental to chatbots like Mitsuku. They have become increasingly effective at processing single sentences, but never grasp the context of the conversations they are having. To improve performance, Steve searches through the mistakes made by Mitsuku in earlier conversations and inserts new, better answers into her database. This leads to improvements in individual answers, but can't improve overall understanding.

It is the quest for true understanding that drives language research at the artificial intelligence labs at both Facebook and Google. While Steve works from the top down – balancing his understanding of the logic of language with insight into which answers work best – the approach of these AI labs is bottom up. Their aim is to train neural networks to learn languages.

The term 'neural network' describes a wide range of algorithms that are inspired by the way the brain works. The human brain is made up of billions of interconnected neurons, which build our thoughts through electrical and chemical signals. Neural networks are a caricature of this biological process. They represent data in the form of a network of interconnected virtual neurons, which take in an input at one end, in the form of a data about the world and produce an output at the other end, in the form of a decision to perform a certain action. For language problems, the input data are the words and the action is to generate the next word in the sequence. Between the inputted words and the outputted actions, the words pass through connections known as hidden neurons. These hidden neurons determine how the input words are converted to output words.

To get a feeling for the bottom-up neural network approach, I looked again at my cat-dog story generator. In its previous incarnation, I took a top-down approach, constructing the logic gates that solved the problem. My bottom-up approach creates a neural network with an input layer that took in the last two words in a sequence. These words were then fed through a hidden layer, the outputs of which are combined in the output layer to produce the next word.

When I first set up my network, the links between the hidden neurons are made at random and the words produced have no structure (Figure 15.2a). 'Dog bites bites chases cat cat dog bites chases dog chases chases dog chases cat …' was a typical output.

The next step is to train the network. The training process involves feeding it words and comparing the output to what I would like it to be. Every time it produces a correct output the links within the network that produce that output are reinforced. So, when it wrote 'cat' after 'dog chases' the links that made this connection are made stronger.[2] By repeatedly strengthening links that give correct sequences, the network takes on a distinctive form (Figure 15.2b) and starts to produce outputs closer and closer to the masterpieces created by my

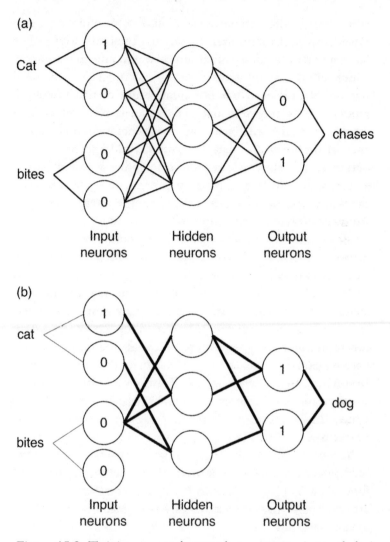

Figure 15.2 Training a neural network to generate 'cat and dog' texts. The input layer of four neurons takes the last two words in the sequence. (a) At first the arrows are weak and random and the output words are generated at random. (b) After training on a sequence of 20,000 pairs of words, some of the links in the network are stronger and others are weaker, depending upon the importance of the input words in predicting the output word.

top-down model. After training on 20,000 sequences, my new algorithm has got the hang of what I want; 'Cat bites dog chases cat bites dog chases cat chases dog chases cat bites dog chases cat chases … ', it says.

In terms of the final outcome, there is no difference between the top-down approach I used at the start of this chapter and the bottom-up approach of neural networks. Both of them cat and dog to infinity. But in terms of how well these models can be generalised to other problems, there is a massive difference. My top-down approach was very specific to cats and dogs, while the bottom-up neural network trained itself based on the rewards it was given.

All I need to create a more proficient neural network author, is to find a very large amount of text on which to train it. This isn't difficult. All of Leo Tolstoy's works are freely available online, and it took me less than a minute to find and download *War and Peace*, as well as *Anna Karenina*. Programming a neural network to read these books and start generating text takes a bit more work, but much of that has already been done too. I asked Alex (of Tinder fame) to see if he could train a network to learn Tolstoy. He soon found a set of programming libraries, created by Google, that allowed him to do the job.[3] It took him just a few days to train them on Tolstoy.

Alex used a subclass of neural networks, known as recurrent neural networks, that are particularly well suited to learning about data that arrives in sequence, as it does when we read the words one after another in *War and Peace*. The input and hidden neurons with a recurrent neural network form a ladder that lifts up the words, combining them to predict a single output word at the top of the network.[4] Alex set up a network that reads in sequences of 25 words and punctuation symbols before trying to predict the 26th. The network reads through the Tolstoy novels many times, trying to predict the words before they come up. When it gets the word right, or predicts a similar word, those connections in the network that led to the correct prediction are reinforced. This is a big increase in complexity from my cat and dog algorithm, which read in just two words, but we have to treat Tolstoy with the respect he deserves.

The technique produces pretty good results. Here is one excerpt:

> *With a message to Pierre's mother, and all last unhappy. She would make the ladies of the moments again. When he came behind him. Behind him struggled under his epaulet. Berg involuntarily had to sympathize.*

I do enjoy the wording here: Berg struck by involuntary sympathy over the unhappy message to Pierre's mother. This is not a phrase used in *War and Peace* or *Anna Karenina*. 'Involuntary sympathy' is an entirely original literary creation by the neural network. Another line that caught my imagination was:

> *Their face almost tickling his head expressed cigarettes, and bulging sacks. It depended on the rhythmic convulsions of the article.*

If we corrected the grammar of 'face' to 'faces' we get a truly menacing picture of a mob of old men with bulging sack-faces surrounding the story's hero. Again, these are phrases not originally used by Tolstoy, but are the neural network's attempt to paste together the type of language he uses.

Recurrent neural networks create text that is quite often grammatically correct, uses punctuation properly and even captures something of the essence of the language of the author they are fed. The output is limited, though. While one or two consecutive sentences might sometimes appear to make sense, longer passages are quickly seen to be nonsense. Here is a slightly longer text:

> *' … At this terrible one of the younger month and threats as still sweet by her, he took their anxiety than asking him, as that? … '*
> *'It is Natásha very much,' asked Natásha, kissed them.*

'Leave one step herself, and still tacked over his attention.
Twice Nicholas had not overstated a bold though he had he felt
very remote as I liked in the father's daughter, still to say, but for
Kutúzov as overcome by expression which yesterday, and only
all sorts of everything the extreme element of Stein aroused by
the young ladies.'

And so it goes on. It soon becomes apparent that reading a
full-length book of this stuff would be impossible: a kind of
James Joyce meets Tolstoy, but without a guiding structure.

While great literature is beyond recurrent neural
networks, it might be possible to train them to handle more
everyday language. At Google, research scientists Oriol
Vinyals and Quoc Le looked at whether they could create an
IT Helpdesk using a recurrent neural network.[5] They took a
database of nearly 100,000 chats between customers who
were facing computer related issues and a specialist who was
trying to help them, and fed the words into a network. After
the training, they had a human ask it some questions. Here
is part of one interaction:

Human: *i am seeing an error related to vpn*
Machine: *what is the error message that you are getting when*
connecting to vpn using network connect ?
Human: *connection refused or something like that*
Machine: *may i know the version of network connect you*
connect ?
Human: *i am not sure i know that*
Machine: *is the network connect prompting for any user*
certificate
Machine: *?*
Human: *yes*
Machine: *are you at home or at work at the moment ?*
Machine: *office ?*
Human: *i am at home*
Machine: *try this*
Machine: *go to URL page and enter your name and password*
Human: *done*

> **Machine:** *try logging in with JohnSmith and check if its working fine now*
> **Human:** *yes, now it works !*
> **Machine:** *great*
> **Machine:** *anything else that i can help*
> **Human:** *no, that was very useful*
> **Machine:** *you are other than JohnSmith :)*
> **Machine:** *have a nice night*

The machine provides service with a smiley.

Many online help services are already automated. Facebook Messenger provides us with shopping bots, either using a series of click-throughs or using simple text messages. Apple's Siri and Amazon's Alexa answer questions and can control devices around your home. These approaches are typically top-down, and thus lack the character of Oriol and Quoc's machine. They also take longer to develop; Oriol told me that his work 'took a couple of months to do, by plugging it in to an existing neural network framework'. Since then, Google has published more of its toolboxes, making it easy for programmers to create their own recurrent neural network chatbots.

The bottom-up bots are not ready for service yet. Sometimes they are less helpful, asking users for pointless information and leading discussions round in circles. Although neural networks might have a bit of personality, most of us want direct help with specific questions when online. When we want information, multiple-choice questions and top-down formulaic answers work best.

Irrespective of whether top-down or bottom-up approaches ultimately prove most effective for creating service bots, one thing is clear: we will find ourselves chatting online with machines more and more in the near future. The Deloitte report on automation in public services, lists customer services as one of the most at-risk jobs. Intelligent chatbots will help us with troubleshooting queries, give us advice on shopping and even provide initial screening for medical problems.

Oriol and Quoc went on to train another neural network to see if they could compete with chatbots like Mitsuku.

They took 62 million sentences from a movie script database and let a recurrent neural network churn through them. The product claimed to be a 40-year-old woman called Julia who came from 'the boonies'. She knew a fair bit about film characters and current affairs, and could even hold a discussion on morality at a level comparable to a stoned student. But she was inconsistent on a number of points, claiming to work both as a lawyer and a doctor, depending on how the question was posed. Even after a joint, most students still know what course they are on.

Oriol and Quoc paid Mechanical Turk workers to compare Julia with a top-down bot, called Cleverbot, to see which approach gave the best answers. Julia scored a narrow victory over Cleverbot. But from what I have read, I'm not sure she would beat Mitsuku. It would be fascinating to see them in a talk-off at the Loebner Prize competition.

Let's get back to reality here. What exactly is Julia? Mitsuku was limited by the time that Steve Worswick could realistically invest answering every possible question someone might pose to his bot. What are Julia's limits? If we fed her a few hundred million more movie scenes, would she become even more realistic?

I discussed these issues with Tomas Mikolov, an authority on computerised language processing. During his PhD studies, Tomas invented the recurrent neural network model that underlies Julia and the Tolstoy generator. He then went on to work for Google, where he created the Word2vec representation of words, now used in everything from web search to translation. Nearly all work on neural network generated language stems from Tomas' research.

Tomas' understanding of the methods has made him sceptical about bots like Julia. He didn't think that Julia was a significant step towards his goal of a true AI. He told me: 'These networks are mostly repeating sentences from the training data and some limited generalisation is achieved through clustering.' The sentences are generated by a human, and the computer repeats them with small changes.

The same critique applies to the Tolstoy generator. The novel sentences that caught my eye are mutant re-combinations of Tolstoy's rich language. Now and again, a nice phrase pops up that the author didn't use originally. Tomas was blunt in his own analysis: 'If you manually hand-pick the generated output, it can look very good and even "intelligent", but this is a totally fake thing to do.'

I agreed. Alex's Tolstoy is fun but fake.

While recurrent neural networks can't have a realistic conversation, they are still revolutionising how we handle text online. Given large enough databases of words, these networks can translate between languages, automatically label scenes in images and provide advanced grammar checks. Once Tomas, Oriol and the other data scientists at Google had set up the basic neural network framework, they were able to provide solutions to translation and labelling challenges that outperformed most of the established top-down approaches. These solutions required little more than feeding lots of words into the networks and letting them learn.

Progress is possible because the hidden layers of a neural network create a way of 'doing maths' on words. In the last chapter, we saw that words can be represented by multi-dimensional vectors. Adding and subtracting these vectors allowed us to test word analogies. Recurrent neural networks provide more complex functions that map between words. The networks find functions that reflect our grammatical rules, our punctuation and identify the important words within a sentence.

By looking inside a neural network trained to translate sentences, Oriol and his colleagues could better understand how the technique works.[6] They found that statements with similar meanings – 'I was given a card by her in the garden', 'In the garden, she gave me a card', 'She was given a card by me in the garden', etc. – generated a similar pattern of activation within the hidden layers of the network. Since the order of the words is very different in these sentences, the hidden layers can be thought of as providing a conceptual understanding. Sentences that mean the same thing are clustered together, with each clustering reflecting a different concept. It is this clustering, which provides a ground 'meaning' of a sentence, through which recurrent neural networks can translate between languages and assign words to images.

When we look deeper inside recurrent neural networks, we can also see the reasons behind their limitations. The problem is not that they need to be fed larger databases of words to gain greater understanding. Their understanding is limited because they can only take in 25 or so words at a time. If we try to train a network with more input words, its conceptual understanding starts to fragment. Neural networks can't convey a concept that takes more than a single sentence to explain, let alone express the ideas that come from actively engaging in a well-written novel or having a good conversation.

Tomas Mikolov is now a research scientist at Facebook Artificial Intelligence Research, and states his overall goal as 'developing intelligent machines capable of learning and communication with people using natural language'. He believes that a deeper conceptual understanding can only be achieved by gradually training a 'learner' bot through a series of increasingly complex tasks.[7] In the beginning, the bot should learn to follow instructions such as 'turn left'. Once it has got the hang of moving, it should be able to 'find food', after which it should be able to generalise to other tasks such as finding information on the Internet. This learning process isn't going to be solved by a recurrent neural network alone, because they lack long-term memory. Tomas has instead proposed a 'road map' of increasingly difficult tasks, that an agent needs to be taught, in order for it to communicate properly with us humans.

Tomas admitted to me that little progress has been made so far. His idea is that researchers should enter 'general AI challenges' each year, where progress is judged on the types of tasks the agents can complete. Rather than relying on superficial human evaluations of how realistic the agents' responses are, these challenges should measure how bots respond to new environments.

Despite Tomas' warning, I couldn't resist doing one last 'totally fake thing' with the Tolstoy neural network. Alex had an idea. He could rewrite *Outnumbered* in the style of Tolstoy. At first I was sceptical – how could that work?

'Simple,' he replied, 'each word in *Outnumbered* is a point in a 50-dimensional space, as is each word of Tolstoy. We can take away Tolstoy words from a machine-generated Tolstoy text, and add the closest words from *Outnumbered*.'

The principle is exactly the same as the one I used in the last chapter to take me from Trump to Merkel, but working in a higher number of dimensions. Alex tried it out. Here are my favourite turns-of-phrase, hand-picked for your enjoyment.[8]

Then all prediction algorithms, felt that lies in an expression about reporting attention, and algorithm up the air of a predict and debate and Donald Hillary of six thousand.

At two people having been statistical that had seemed to himself his wife's chances that however much these online had only more desire to website out his feelings for both browsers' outcome.

His notes and reality as much as his result, did not so in the possibility of polls. 'At once myself because way,' said the question, 'commit the young woman!' Only thought between the decimal.

Maybe I can let the algorithm finish off the book? Well, maybe not. There still some serious questions that needed to be answered, not least about our journey towards artificial intelligence. I wanted to know where we were on Tomas' road map. How far is artificial intelligence from catching up with us humans?

CHAPTER SIXTEEN

Kick Your Ass at *Space Invaders*

I have never been particularly interested in chess. I'm sure a good part of my ambivalence towards the game is grounded in my inability to understand it. I know the rules, but when I look at the pieces, nothing happens in my brain. I can't tell a good move from a bad move, and I can't see more than one or two steps ahead. The game is a mystery to me.

So, for me, a computer beating Garry Kasparov in 1997 was not particularly exciting. It was just slightly curious that it hadn't happened sooner. Computers can calculate many more chess moves per second than a human, and it seemed inevitable that they would eventually triumph over us. When IBM's supercomputer Deep Blue completed the task, I was far from surprised. The algorithm had been fed 700,000 grandmaster games and could evaluate 200 million positions per second. It beat Kasparov with brute force, accumulating more data than he could possibly store and making more calculations than he could possibly make.

The defeat of Kasparov was not widely considered as an important step towards a more general artificial intelligence. At the time of Deep Blue's victory, the field of AI had gone out of fashion. While a computer could win at chess, it was proving difficult to get a robot arm to pick up a cup of water. Change the position of the cup or put a mug with a handle in its place and even the most advanced robots spilt the water everywhere. When I studied computer science as an undergraduate at Edinburgh University in the early 1990s, it was possible to do a joint degree incorporating artificial intelligence. My course advisor told me the subject was a dead-end and that I should combine my degree with statistics instead. He was right. The top-down AI of the 1990s slowly faded away and statistics took its place as the tool behind modern algorithms.

Sheer computing power has beaten us at more and more games. When in January 2017, an algorithm called Libratus played 120,000 hands of heads-up, no-limit Texas hold 'em against four leading professionals and won, the victory had again been achieved by brute-force calculation. The Carnegie Mellon University scientists who designed the algorithm found the probability of winning every potential type of hand.[1] Played at the highest level, poker is not a game of psychology. It is simply a question of knowing the odds. After 25 million processor hours, Libratus knew the odds better than any human possibly could. It slowly, but surely, whittled down its opponents' pile of chips.

Impressive as they are, achievements in board and card games require the programmers to tell the algorithms how to solve the problem using a top-down solution. They don't qualify as contributions to finding a more general bottom-up form of artificial intelligence.

Space Invaders, though. That is a whole different game. The Atari classic might not be as intellectually demanding, but it combines strategic planning with hand-eye coordination and quick reactions. I also happen to enjoy playing it and I was a pretty decent player when I was nine or 10. It is a game that most of us can relate to. Although it is played on a computer, it is a very human game.

So when, in 2015, the Google DeepMind research team published an article in the journal *Nature* showing that a computer could learn to play *Space Invaders* to the same level as pro gamers, I was duly impressed.[2] The big difference between Google's *Space Invaders* algorithm and IBM's chess algorithm was that the former had taught itself to play. Just as I had sat down in front of the TV screen, switched on the Atari 2600 console my family had borrowed from friends, and spent the next weeks playing the game for as long as I was allowed to, so too had Google's neural network learnt the game by playing. Plugged into a computer screen and a joystick, the neural network had played the game over and over again, very badly at first, but slowly improving. Until, after the equivalent of 38 days of game time, it was better than I had ever been. In fact, it was about 20 per cent better than a professional human games tester.

Watching how Google's neural network plays the game takes me back to the early 1980s. It positions the tank behind the barriers, picks out one line of invaders at a time and takes down the spacecraft for bonus points. Finally, it carefully positions itself, as the final line of aliens accelerates towards earth, eliminating them one by one. Atari aficionados will be interested to hear that the algorithm does not use the tactic of making a narrow slit through one of the houses and shooting from there, a method decreed as 'cheating' in our household. Instead, the computer relies on accurate shooting to pick off the aliens.

The Google team didn't stop at *Space Invaders*. They set up a neural network to learn to play 49 different games on the Atari 2600 console. Out of these it beat pro gamers in 23 games, and achieved a level equivalent to the average gamer in a further six games. It was especially good at *Breakout*, a game in which you control a paddle and try to knock out all of the blocks from a wall. After the equivalent of a week of non-stop game play, it learnt the tunnelling strategy, where the player opens up a small hole on the side of the blocks and sends the ball through to the top of the brick wall. Once this opening is made, the ball bounces around on the top of the wall, quickly demolishing the bricks. When my friends and I discovered this strategy, we declared the game to be boring. Not so for Google's neural network. It continued to play and play, racking up scores that far exceeded those that any human could achieve.

Before the neural network can successfully play a game, the connections between the hidden neurons need to be tuned in order to get the correct set of outputs for each input. If the aliens are above the spaceship, we want the algorithm to press the button to shoot. If an alien bullet is about to land on top of the spaceship, the algorithm should move it behind a barrier. It is here that training comes in. Initially, the Google engineers told their neural network algorithm absolutely nothing about the game it was about to play. They set up the network with random connections between the neurons, which means that the spaceship moves and shoots more or less at random.

With random settings, the neural network loses a lot of games, but now and again it 'accidently' shoots an alien and gets points. The training process looks through this long list of screenshots (inputs), joystick moves (actions) and scores (outcomes) and works out whether the actions performed increased or decreased the neural network's game score. The network is then updated so that connections that give points are strengthened and those that result in loss of a life are weakened. After weeks of training on some of the fastest computers in the world, the neural network can link specific screen patterns to the joystick actions that give the most points.

The exact same approach was able to learn to play games as varied as *Robotank* (an early '3D' shoot-'em-up game), *Q*bert* (a puzzle platform game), *Boxing* (a top-down boxing game) and *Road Runner* (a sideways scroller). The on-screen patterns in these games are very different: in *Robotank* you chase after enemy tanks and try to shoot them; in *Q*bert* you control a little orange creature that tries to paint hexagons and avoid being caught by a purple snake; in *Boxing* you punch an opponent in the face and in *Road Runner* you run along the road trying not to get knocked down or caught by a fox. Over many rounds of playing each game, the neural network slowly grasps the underlying pattern and works out which objects of which size are most important in finding a winning strategy. The Googlers had created an artificial intelligence that learnt each game from scratch.

Humans find the task of identifying patterns within a game very easy. When the nine-year-old me saw *Space Invaders* for the first time, I could immediately distinguish the rows of aliens, the houses and the defending tank. But this exact same challenge of finding the pattern that defines a game had, before the DeepMind Googlers tackled the task, proven too much of an obstacle to previous attempts to get computers to learn to play games. The computer couldn't work out what the games were about.

An important part of the solution was the use of a mathematical technique called 'convolution'. We often use the word 'convoluted' in connection with a 'convoluted explanation', meaning a long, detailed story that is difficult to follow. In the case of the neural network, the thing that is convoluted is the screenshot of the game. When playing *Space Invaders*, the input to the neural network is the 210 x 160-pixel screen display of the Atari 2600. In the first hidden layer of the neural network, the original screenshot is stretched out as a series of smaller pictures that are then input into the hidden neurons in the network (Figure 16.1). This process is repeated on the resulting images, within a second and then a third network layer, to produce even smaller images in deeper hidden neurons. At this point, the original screenshot is highly convoluted: it is now represented by lots of small images that each capture only a small part of the overall

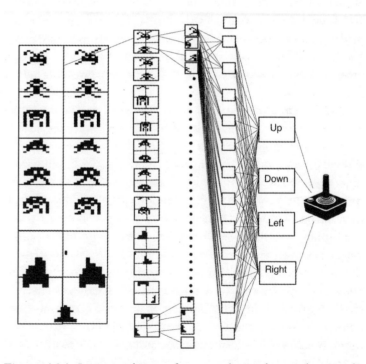

Figure 16.1 Inner workings of a convolutional neural network. Figure drawn by Elise Sumpter.

picture. Each of these images is repetitive and difficult to put in a wider context: a bit like that never-ending story your uncle Jim tells you about his schooldays.

The deeper layers of the neural network put the small pictures back together again by feeding them through a fourth and fifth layer of fully connected neurons (Figure 16.1). It is within these layers that the network learns a relationship between the small images and the best action to perform. It is also here that it works out the size of the important patterns in the game. If the network is learning about a game where the important objects are large, like the boxers in *Boxing*, then the network will form lots of very similar connections between the neurons that are near to each other. If the objects are small, like space invaders or the hexagons that *Q*bert* aims to paint, then the connections will be more intricate. If the objects can change size, like the tanks as they become closer in *Robotank*, then the connections will be similar between different convolution layers. Convolutional neural networks are a powerful technique because they find the size and shape of important patterns automatically, without programmers having to tell the neural network what to look for.

The idea of convolutional neural networks has been around since the 1990s, but for a long time, they were just one of many different, competing algorithms that had been proposed to help computers detect patterns. It was in 2012, when Alex Krizhevsky gave a four-minute presentation at the Neural Information Processing Systems conference at Lake Tahoe, California, that researchers began to take special notice of the convolutional method. Alex had entered a competition, run yearly as part of the conference, to automatically identify different objects in pictures.[3] For a human, this is a mundane task. The pictures are of a variety of scenes: some men showing off the large fish they have caught; a row of sports cars; a bar full of people and two women taking a selfie, are just a few of a whole range of scenes depicting many different aspects of life. The task is to find the objects – the fish, the men, the cars, the people in the bar and the phone used to take the selfie. Before Alex presented his work at the conference, this image-identification task had proven very difficult. Even algorithms that had been fine-tuned to the problem, made errors at least one time out of four.

Alex reduced this error rate to one in six using a convolutional neural network. He didn't tell the algorithm much about the size and shape of the objects it was asked to classify. He just let it learn. And when it was fed millions of images it learnt very well indeed. While other approaches relied on humans to work out the important features – such as edges, shapes and colour contrasts – that define objects, Alex's approach was to just set up the network and let it do the work.

The 2012 competition entry was just a start. Alex was a PhD student using a couple of gaming graphics cards installed on his computer. Once the technique was out, other researchers refined it and put more powerful computers onto the job. The next year, the winning entry made mistakes in only one case out of eight, and by 2017 the error rate was below two per cent.[4] Both Google and Facebook started to take notice. They realised that convolutional neural networks solved problems at the heart of their businesses. An algorithm that can automatically recognise our friends' faces, our favourite cute animals and the exotic places that we have visited can allow these companies to better target our interests.

Alex and his PhD supervisor, Geoffrey Hinton, were recruited by Google. The following year, one of the competition winners, Rob Fergus, was offered a position at Facebook. In 2014, Google put together its own winning team, and promptly recruited Oxford PhD student Karen Simonyan, who came in second place. In 2015, it was Microsoft researcher Kaiming He and his colleagues who took the prize. Kaiming was recruited by Facebook the next year.[5] One by one, the leading neural network researchers were whisked away to work at the newly formed artificial intelligence groups at Microsoft, Google and Facebook.

These researchers weren't just recruited in order to find objects in images. The important breakthrough of Alex's research was that it demonstrated that convolutional neural networks could learn to solve a problem without having to be 'told' which problem it was they were solving. It soon became apparent that the same approach beat its competitors on handwriting and speech-recognition tasks. It could be used to recognise actions in short films and predict what was going to happen next.

This was why convolutional neural networks that played Atari games were so much more exciting than the algorithm that had beaten Kasparov at chess. When Deep Blue won, the researchers had established that a computer could beat a human at an esoteric game. When the contest was over and the media had done their interviews, the machine was switched off and the researchers went back to their everyday jobs.

The arrival of neural networks was different. The algorithms solved one problem after another. Face recognition in Apple's iPhone X uses neural networks to uniquely identify its owner's face. Tesla uses neural networks in its car vision system to warn about potential collisions.[6] The achievements of convolutional neural networks on vision problems were accompanied by similar improvements by recurrent networks on language problems. Google has made big improvements to the quality of its translations from English to Chinese using new neural network techniques.[7]

The potential scale for improvements in algorithms based on neural networks remains unclear. At the very least, recent research has vastly increased computers' abilities to recognise objects, sounds and sentences. But the excitement around this method has led to suggestions that we are on our way to even more dramatic results. Are we finally close to solving the problem of creating generally intelligent machines?

I certainly wanted to answer this general AI question. But before I could get near to solving such a big question, I felt I needed to answer a smaller question first. How good were convolutional neural networks? Alex Krizhevsky's image analysis paper had revolutionised both industry and academia, but I wanted to better understand the limits of neural networks.

KICK YOUR ASS AT *SPACE INVADERS*   219

Despite all the hype, a large part of the answer can already be found in Google's own research. Let's go back to the Atari game-playing neural network. The researchers found that it could play 29 out of 49 games they tested at human level, beating the pros at most of them. This means that humans are better than neural networks for 20 of the tested games. For some of these, the neural network's performance was little better than random.

When I looked at the list of computer-vs-human performances, one game, in particular, caught my eye: *Ms Pac-Man*. The performance of the neural network on this simple maze game, in which *Ms Pac-Man* has to munch as many food pellets as possible while avoiding being caught by ghosts, was very poor. It managed scores of around 12 per cent of those achieved by a pro gamer.

Ms Pac-Man is trickier than *Breakout* or *Space Invaders* because it involves patience. If Ms Pac-Man is going to survive, then she should be cautious, waiting until an area doesn't contain ghosts before moving there. The 'power pellets' that turn the ghosts blue and allow Ms Pac-Man to eat them, should be used sparingly. If she takes them early in the game, then Ms Pac-Man will get a few points from eating ghosts, but will find it difficult later on when the ghosts regenerate and hunt her down, four against one.

Convolutional neural networks simply can't deal with these aspects of game play. They can only react to what they see directly in front of them: aliens that need to be shot, a boxer that needs to be punched and hexagons that need to be hopped on. They can't plan, even for a very short-term future. The algorithm fails on all Atari games that require even the slightest degree of forward planning.

The structure of these networks means they are good at identifying objects in pictures, putting together sounds to make up words and recognising what to do next in a shoot-'em-up game. But we can't expect them to go beyond these tasks, and nor do they. Today's most advanced artificial intelligence can see things and produce immediate responses, but it can't understand what it is looking at. It can't make a plan.

I wasn't the only person whose attention had been caught by *Ms Pac-Man*. Microsoft researcher Harm van Seijen told me that, after reading the Google paper, the game caught his eye. He was surprised that the neural network hadn't solved the game and wondered why it differed from *Space Invaders*.

He and his colleagues developed an alternative approach. They worked out that the *Ms Pac-Man* problem was most easily solved when broken down into smaller problems. They modelled all the components of the game – the pellets, the fruits and the ghosts – as agents that competed for Ms Pac-Man's attention. They then trained the neural network to balance the strength with which these agents 'pulled' on her at different points in the maze. The result was a very cautious Ms Pac-Man, who took a longer time to complete levels but never got eaten. Their algorithm racked up a perfect score of 999,900, at which point the game crashed.

The researchers developing neural networks are acutely aware of the problem of 'telling their algorithms too much'. The long-term goal of the general artificial intelligence programme is to train networks with as little human input as possible. If we want algorithms to exhibit the qualities we see in animal or human intelligence, then these networks must learn for themselves, without us telling them what to do or what patterns to pay attention to.

On the other hand, these same researchers are keen to demonstrate that they can win at more complex, modern computer games. The ultimate challenge is *StarCraft*, a sophisticated strategy video game played competitively as an e-sport. Google's DeepMind team has, together with the games creator, Blizzard, set up a software environment to help researchers build an algorithm to learn to play *StarCraft II*.[8] This environment provides an abstract representation of the game, rather than the screen view shown to human players, allowing programmers to skip the challenge of recognising the objects on the screen.

Harm was very sceptical about the possibility of learning modern computer games from the on-screen pixel view without providing the algorithm with additional information.

He told me: 'Learning *StarCraft* from scratch? That's not going to happen.' He said that learning *StarCraft* would require the programmers to provide specific game information and to use a highly specialised neural network.

Harm's own approach to *Ms Pac-Man* lies somewhere between the pure neural network approach to the 'easier' Atari games, and the work on *StarCraft*. Unlike in the neural network that plays *Space Invaders*, Harm told his network the position of *Ms Pac-Man*, the ghosts and the pellets. But he didn't tell the network in advance that pellets were good or that ghosts were bad. Instead, the network learnt how to balance the risk of moving towards a ghost, with the benefits of moving towards food. It then learnt how to put this information together.

When I talked to Harm, he kept returning to this challenge of balancing how much information he should give his neural network. 'I'm interested in how to get computers to learn behaviour by interacting with the world,' he told me, 'I don't specify what an agent [like Ms Pac-Man] needs to do, I only specify what she wants to achieve.' For Harm, this is the fundamental challenge for research, rather than simply getting a computer to get a high-score at *Ms Pac-Man*.

This challenge of finding out how much a computer can learn from scratch, is central to understanding how far we are from creating a general AI. After I talked to Harm, in October 2017 I contacted David Silver, who leads the DeepMind team that is training neural networks to play the board game Go.

The AlphaGo algorithm that David's team created had beaten the world number-one Go player, Ke Jie in May 2017. The algorithm had begun life by learning a playbook of 30 million moves made by the world's best Go players. It then fine-tuned its skills by repeatedly playing against different variations of itself. The final algorithm combined an encyclopaedia of game knowledge, a neural network that had learnt through playing and a powerful calculating machine that searched through potential moves. It is an impressive engineering feat, but it is a highly specialised algorithm.

David had also been involved with the Atari games project, so I felt that he would have insight into the balance between learning from scratch and building a specialised algorithm, as he had done with Go. I emailed him a series of questions about this, but he replied asking me to be patient because a 'new paper in a few weeks' would answer my questions.

It was worth the wait. On 19 October 2017, David and his team published an article in the journal *Nature* describing AlphaGo Zero, a new Go-playing algorithm that beat all previous algorithms. Not only that, this algorithm worked without human assistance. They set up a neural network, let it play lots of games of Go against itself and a few days later it was the best Go player in the world.

I was impressed, much more so than when a computer won at chess or poker, or even with David's first Go champion. This time, there was no rulebook or specialised search algorithm, just a machine playing game after game, from novice up to master level, up to a level where nobody, computer or person, could beat it. It was a neural network learning to play a very complex game all by itself.

David told me he didn't think there was any reason his approach couldn't learn to play *Ms Pac-Man* too, although Google hadn't yet tried. For David, the new Go champion answered many of the doubts about how widely DeepMind's neural networks could learn to solve different problems from scratch. He and his colleagues stressed that, 'told the rules of the game, [the neural network] learns by trial and error'.

When I asked Harm about the new Go champion he told me this was 'definitely impressive and a clear improvement', but he wasn't convinced that this was learning from scratch.

Harm told me that 'the rules of the Atari games are unknown to the algorithm and have to be learnt'. The challenge in learning to play computer games from the bottom up is working out how the screen changes in response to an action. This information is already provided to the computer for Go.

Tomas Mikolov, the Facebook engineer who developed language-processing neural networks, didn't see Go as a step towards a more general AI. 'There is a danger of over-optimising something irrelevant for the final goal [of creating AI] if one works with highly artificial problems like computer games or Go, chess, etc.,' he told me. Tomas believes that it is only by working with language-based tasks, where we communicate and teach an AI system, that we can develop true intelligence.

I thought back to my own childhood. When I saw *Ms Pac-Man* or *Space Invaders* for the first time, it took me no more than a few minutes to understand what was going on. It didn't seem strange or unnatural to be able to control the figures on the TV and my brain quickly worked out the connection between the joystick and the screen. Pretty soon I was racking up high scores for one game cartridge after another. I witnessed the same thing when my 12-year-old son opened up Blizzard's *Overwatch* for the first time on the PlayStation 4. His mind immediately made sense of the moving figures, understood how the controls worked and started to think about the strategic elements of the game.

When children learn to play computer games they use a very different strategy than that adopted inside a neural network. They come to the game with an established model of how objects interact, both in real life and within games. Brenden Lake at New York University, working with colleagues at MIT and Harvard, looked more closely at an Atari game called *Frostbite*. They noticed that they could learn to play the game much more quickly than a neural network, since they quickly grasped the goals of the game and the way the objects moved. It took them just two minutes of watching a YouTube video of someone playing the game, plus 15–20 minutes game time to get their score up to the same level as a neural network.

Brenden also posed some interesting thought experiments. He noted that he could easily adapt his style of play to, for example, get as many fish as he could in the game or go as long as possible without dying. For the neural network, completing these tasks would mean starting its training again from scratch. Although scientists understand a lot about the brain, we remain spectacularly incapable of simulating humans' spontaneous understanding of new contexts.

Google, Tesla, Amazon, Facebook and Microsoft are all in a race to produce this understanding. Much of their work is collaborative; they share libraries of code and share the latest knowledge at conferences, like NIPS. There is a real buzz about the progress being made and a friendly rivalry between the groups. Some of these engineers and mathematicians believe we are only a decade or so from true AI. Others see it as centuries away.

I wondered what their bosses made of all the excitement.

CHAPTER SEVENTEEN

The Bacterial Brain

In 2016, Facebook CEO Mark Zuckerberg set himself a challenge: to build an automated butler that would help him around the house. The name of his butler, Jarvis, was inspired by an AI robot built by the character Iron Man from the comic and film series *The Avengers*. The fictional Jarvis has a human-like intelligence, reading Iron Man's thoughts and connecting emotionally with him. Jarvis combines a massive database of information with the power to reason and grasp concepts. Mark didn't need his assistant to help him save the world, but he did want to see just how intelligent a home help he could create using Facebook's library of algorithms.

Google has ambitions that stretch beyond personal butlers. The DeepMind team, which had won at Go and *Space Invaders*, has helped Google improve energy efficiency of its servers and developed more realistic speech for the company's personal assistant. Another application area is DeepMind Health, a project one of the London Googlers had told me about when I visited them. The aim is to look at how the National Health Service in the UK collects and manages patient data in order to see how the process can be improved. DeepMind's CEO Demis Hassabis talks about his team one day producing a 'high-quality scientific paper where the first author is an AI'. The ultimate goal is to replace many of the difficult intellectual challenges carried out by engineers, doctors and scientists, with solutions created by intelligent machines.

The researchers working on these projects often claim they are making small steps towards a more general artificial intelligence. Many of them believe that they are taking us on a journey ever closer to the so-called 'singularity', the point at which computers become as smart as us. The singularity hypothesis says that once this point is reached, once computers are able to design other intelligent machines and systematically improve themselves, then our society will change dramatically and forever. The machines might even consider us surplus to requirements.

At a meeting of the Future of Life Institute – a charitable organisation in Boston, Massachusetts, focused on dealing with future risks – in January 2017, theoretical physicist Max Tegmark hosted a panel debate about general artificial intelligence.[1] The panel included nine of the most influential men in the field, including entrepreneur and Tesla CEO Elon Musk; the Google guru Ray Kurzweil; DeepMind's founder Demis Hassabis and Nick Bostrom, the philosopher who has mapped our way to, what he calls, 'superintelligence'.

The panel members varied in their views as to whether human-level machine intelligence would come gradually or all of a sudden, or whether it will be good or bad for humanity. But they all agreed that a general form of AI was more or less inevitable. They also thought that it was sufficiently close that we needed to start thinking now about how we would deal with it.

Despite the panel's conviction that AI is on its way, my scepticism increased as I watched them talk. I had spent the last year of my life dissecting algorithms used within the companies these guys lead and, from what I have seen, I simply couldn't understand where they think this intelligence is going to come from. I had found very little in the algorithms they are developing to suggest that human-like intelligence is on its way.

As far as I could see, this panel, consisting of a who's-who of the tech industry, wasn't taking the question seriously. They were enjoying the speculation, but it wasn't science. It was pure entertainment.

The problem with my position – that talking about what will happen when general AI arrives is idle speculation – is that it is difficult for me to prove. Part of me feels like I shouldn't even try. If I do, I am just joining in the clamour of middle-aged men desperate to have their opinion heard. But it seems that I can't help myself. With Stephen Hawking claiming that AI 'could spell the end of the human race', I can't help wanting to clarify my own position.

I have tried to argue against the likelihood of general AI before. In 2013, I had an online debate with Olle Häggström, a professor at Gothenburg University, about the subject.[2] Olle believes that the risk is sufficiently large that we should make sure that humanity is prepared for its arrival. We have to make sure we minimise the risks of any potential transition to superintelligence.

My counter-argument to excessive preparation for AI now is that it is just one of all sorts of unknown risks in the future. We are still struggling with global warming. We live in an age where a nuclear war could start with a few hours notice. If we think ahead 100 years, it is a realistic possibility that our planet is hit by a large meteor or a massive solar flare, that we experience a volcano that darkens the sky for many years and/or that we enter a severe ice age. These are all challenges we know about and need to prepare for.

There are many other threats that can arise from technology. Imagine if biologists accidentally (or otherwise) genetically engineered a supervirus or a deadly bacteria strain that infects and slowly kills all mammals, including us.

Imagine what would happen if we found a way of prolonging human life indefinitely so that no one needs to die. The demand on resources would be immense and conflicts inevitable. Think about the risks that might be created by the emergence of a grey goo of nanoparticles. This goo could be created by scientists on a very small scale, developing the ability to reproduce and 'eating' everything on our planet.

If we are going to go as science fiction as general AI, then we also need to consider how we might prepare for the discovery of alien intelligence when the James Webb Space Telescope starts taking more detailed pictures of our universe. What do we do if we find that the stars are moving in a way that contradicts the rules of physics and can only be explained by extraterrestrial intelligence? And what about the theory in

The Matrix that we are all living in a computer simulation? Shouldn't we invest more research in checking for potential anomalies in our reality?

If any of these things happen before general AI, then they pose just as great a risk to humanity as a superintelligent computer. Ironically, many of these doomsday scenarios are not my own, but have been carefully documented by Olle. I found them through our discussions and reading his excellent book *Here Be Dragons*.[3]

Olle isn't convinced by my argument though. He still considers general AI to be one of the bigger risks to our continued existence or, at the very least, a risk that should be carefully controlled.

I decided I'd try a different approach this time. An approach that is less philosophical and more practical. I'll concentrate on giving a clear idea of where we are now. From there, I'll let you (and Olle) draw your own conclusions about what the future holds.

In another discussion at the same conference as Max Tegmark's panel debate, Yan Le Cun, inventor of convolutional neural networks and Facebook's lead AI researcher, described how solving the image-identification problem, using his method, had been like climbing over one mountain.[4] Now they were over the top and back in a valley looking up at the next peak. Yan didn't know how many more mountains there were to climb, but thought there could be 'another 50'. DeepMind's Demis Hassabis put the number of mountains at less than 20. In his view, these mountains each comprise a list of unsolved problems about how we simulate different, known properties of the brain.[5]

The mountain metaphor raises more questions than it answers. How do they know from looking at the next mountain if it can be conquered? They don't have a map of the territory or a clear route to take them up to each summit. And worse still, how do they know that they won't climb over one of the mountains and find a peak that simply can't be ascended? What does climbing over one mountain tell us about the next?

A good way of putting recent algorithmic development in context is to think about the tasks these algorithms can and can't currently do, and why. During the question-and-answer session after the Future of Life Institute discussion, Anca Dragan, assistant professor of electrical engineering and computer science at Berkeley, who was generally more sceptical about AI developments than many of the other participants, gave an example of an apparently simple challenge that was unlikely to be solved in the near future. She said she wouldn't be surprised if robots would be unable to empty a dishwasher stacked by a human for a few years to come.

Another sceptic, Oren Etzioni, CEO at the Allen Institute for Artificial Intelligence, took up language as an example. He told the audience that computers 'can't reliably determine what the word "it" binds to in a sentence. And *it's* pretty damning for people who believe that [computers are] about to take over the world.'

We know that the 'it' in his second sentence refers to the overall implication of the first sentence. Even the best language algorithms are unable to determine if his 'it' means a sentence, a computer, the word 'it' or another idea within the context of the sentence.

My own favourite example of what AI can't do comes from football. Highlights of recent matches can be found online. Two robots repeatedly walking into each other while the ball lies stationary half a metre away from them; a robot goalkeeper failing to dive as a softly stuck shot rolls slowly past it; players repeatedly falling over as the force of their own kicking motion propels them downwards. Watching these robots illustrates just how far we have to go.

After the 2016 world RoboCup competition, I interviewed Tim Laue and Katie Genter who were, respectively, members of the winning team from Bremen, Germany, and the runner-up team from Austin, Texas, in the standard robot class. They told me that they focus on writing algorithms that detect the lines of the pitch, the goalposts, the ball and the other players. Each algorithm used is problem-specific: performing ball detection or kicking motion. This top-down approach is a long way away from the bottom–up AI that would ultimately be required if robots are to beat human opponents. The robots are performing a sequence of identification tasks, rather than learning how to play the game. There is still a very long way to go until we have an AI football player.

If AI can't do human–level tasks yet, maybe it can compete with animals? Can we scale down our ambitions slightly and create an algorithm that is as smart as a dog?

Some people, including many scientists who should know better, talk about animals in terms of simple stimulus-response reactions. The classic example is Pavlov's dog salivating at the sound of a bell. Anyone who owns a dog will tell you that this Pavlovian view is a vast over-simplification, and they are right. A typical dog-owner's view of their pets as family members and friends is not just an emotional, anthropocentric view. It is in line with how most modern behavioural biologists see domesticated animals – as sharing many of our complex behaviours. Juliane Kaminski, head of the dog cognition project at the University of Southampton, has found that dogs can learn in a similar way to small children, take into account their owner's perspective of the world when deciding which objects to fetch, and understand our intentions from our body movements.[6]

These qualities, of understanding the context of different situations and learning about how to learn, remain open questions in AI research. Until we have made much more significant progress towards modelling a human than we have up to now, we won't be able to simulate dogs, cats or other domestic animals.

I have maybe set my sights too high with dogs, so let's skip down a few levels to insects, and bees in particular. Lars Chittka, at Queen Mary University of London, has recently documented our increased understanding of bee cognition and found bees have an amazing intellect.[7] After a few flights looping around their nests, newborn bees have a good idea of what their world looks like. They then quickly set to work collecting food. Worker bees learn the smell and colour of the best flowers and solve the 'travelling salesman problem' of visiting the available food sources in the shortest possible time. They can remember where they have experienced threats and sometimes 'see ghosts' as they react to a perceived danger that isn't there. Bees that find lots of food become optimistic and start to underestimate the danger of predator attacks. The underlying neural network, in the form of the bee's brain, has a very different structure to that of artificial convolutional or recurrent neural networks. Bees seem to be able to recognise the difference between objects using just four input neurons, and appear to lack any internal representation of images. Other simpler stimulus–response tasks, which could be modelled as a small number of logic gates, instead affect entire regions of the brain.

The most remarkable thing about bees is that they can learn to play football! Well, not quite football, but a game very like it.[8] Lars's research group has trained bees to push a ball through a goal. The bees could learn to do this task in a variety of different ways, including watching a plastic model bee push the ball and watching other real bees complete the task. They didn't need to extensively practise the game themselves in order to learn the task. Ball-rolling is not something bees usually encounter in their lives, so the study shows that bees can learn novel behaviours quickly, without the need for repeated trial-and-error attempts. This is exactly the problem that artificial neural networks have failed to overcome so far. Bees can generalise their skills in other areas to tackle a new problem like football.

It is important to remember here that the question of general artificial intelligence isn't about whether or not computers are better at particular tasks than humans. We have already seen that a computer can play games like chess, Go and poker better than humans. So I don't think the machines would have much problem beating a bee at these games. The question is whether we can produce bottom-up learning on a computer of the type widely observed in animals. For now, bees are able to generalise their understanding of the world in a way computers are not.

The nematode worm *C. elegans* is one of the simplest living animals. A fully developed adult consists of 959 cells, of which around 300 are neurons. This compares with the 37,200,000,000,000 cells in your body[9] and 86,000,000,000 neurons in your brain.[10] *C. elegans* are widely studied because, despite their relative simplicity, they share many of our properties, including behaviour, social interactions and learning.

Monika Scholz at the University of Chicago has recently created a model of how the worm uses probabilistic inference, similar to that used in Nate Silver's model of election polls in Chapter 8, to decide when to move.[11] The worm 'polls' its local environment to measure how much food is available and then 'predicts' whether it is better to stay put or start exploring for new resources. Studies like these reveal details of worm decision-making, but they don't yet model the organism as a whole. A separate project, known as OpenWorm, attempts to capture aspects of the mechanics of how worms move, but more work is needed to put these models together and reproduce *C. elegans*' full repertoire of behaviour. For now, we don't really know how the 959 cells act together and thus can't properly model the behaviour of one of the simplest animals on Earth.

So let's forget, for the time being at least, about creating the intelligence of a dog, a bee, a worm or even a football player. How about an amoeba? Can we reproduce the intelligence of microorganisms?

The slime mould, *Physarum polycephulum*, is an amoeboid organism that builds tiny networks of tubes to transport nutrients between different parts of its body. Audrey Dussutour at Toulouse University in France has shown that slime moulds habituate to caffeine, a substance they usually try to avoid, and then revert to their normal behaviour when given an option of avoiding the substance.[12] Other studies have shown that slime moulds can anticipate periodic events, choose a balanced diet, navigate around traps and build networks that efficiently connect up different food sources. The slime can be thought of as a form of distributed computer, taking in signals from different parts of its body and making decisions based on its previous experience. It does all of this without a brain or a nervous system.

It may be possible to produce a comprehensive mathematical model of slime moulds in the near future, but we are certainly not there yet. The 'memory' and learning of slime could potentially be modelled by a type of electrical component known as a 'memristor', a combination between a resistor and capacitor, that provides a form of flexible memory.[13] But we still don't know how to set up a network of memristors to combine with each other and solve problems in a way that mimics the slime mould.

The next step down in biological complexity from slime moulds are bacteria. The bacteria *E. coli* is a 'bug' that lives in our gut. Although most strains are benign or even beneficial, a few of them give us food poisoning. *E. coli* and other bacteria navigate through our bodies, take in sugars and 'decide' how to grow and when to split.[14] They are highly adaptable. When you drink a glass of milk, the genes for lactose uptake are activated within *E. coli,* but if you then eat a chocolate bar the genes that process glucose, which the *E. coli* 'prefers', suppress the lactose genes. Bacteria move around through a run-and-tumble motion, making runs in one direction before tumbling to 'choose' a new direction. They tune these 'tumbling' rates to the quality of the environment they find themselves in. Each of the bacteria's different 'objectives' – obtaining resources, moving around and reproducing – are balanced through different combinations of genes switching on and off.

Does *E. coli*'s balancing of different objectives in a quest to obtain resources sound familiar? It should do. Ms Pac-Man is a bacteria. The tasks these two organisms, real and artificial, aim to complete are very similar. In order to adapt, they both have to respond to the input signals from a variety of different sources: *E. coli* regulates intake of resources, responds to dangers and navigates obstacles. Ms Pac-Man's neurons respond to ghosts, food pellets and the structure of the maze. The bodies that bacteria live in are not identical, just like *Ms*

Pac-Man mazes differ from each other, but the algorithms each of them employ are flexible enough to handle a wide range of environmental challenges.

I had found the closest biological equivalent of the highest level of current AI. It is a tummy bug.

One argument against my bacteria-brain analogy, is that the reason we can't simulate worms and slime moulds is that we don't know what these organisms are aiming to achieve. Some of the neural network researchers I talked to argued that we don't know what these experts call the objective function of worms. To train a neural network we need to be able to tell it what pattern it is meant to produce and, in theory, if we know the pattern, *i.e.* the objective function, we should be able to reproduce the pattern. There is some validity to this argument – biologists don't have a full understanding of *C. elegans* or slime moulds.

Ultimately, however, the 'tell us the objective function' argument sidesteps the real issue. Biologists' experimental work on intelligence reveals more about *how* the brain works: understanding the connections between neurons and the roles of different parts of the brain, than it reveals about the overall pattern of *why* our brains have particular objectives. If neuroscientists are going to work together with artificial intelligence experts to create intelligent machines, then this joint work can't rely on biologists finding the objective function of animals and telling it to the machine-learning experts. Progress in AI must involve biologists and computer scientists working together to understand the details of the brain.

Tests of AI should, in my view, build on the one first proposed by Alan Turing in his famous 'imitation game' test.[15] A computer passes the Turing test, or imitation game if it can fool a human, during a question-and-answer session, into believing that it is, in fact, a human. This is a tough test and we are a long way from achieving this, but we can use the main Turing test as a starting point for a series of simpler tests.

In a less well-cited section of his article from 1950, Turing proposes simulating a child as a step toward simulating an adult. We could consider ourselves having 'passed' a mini imitation game test when we are convinced the computer is a child. My argument is that we should use the rich diversity of organisms on our planet as a series of test cases.[16] Can we reproduce the intelligence exhibited by slime mould, worms and bees in a computer model? If we can capture their behaviour when they're moving around their environments and interacting with each other, then we can claim to have produced a model of their general intelligence. Until we produce these models, then we should be careful about the claims we make. Based on current evidence, we are modelling intelligence on a level similar to that of a single bacterium.

Well … not quite. Harm van Seijen had been very careful to explain that his *Ms Pac-Man* algorithm could not be considered as having been built from scratch. He had helped it by telling it to pay attention to ghosts and pellets. In contrast, the bacteria's knowledge of the dangers and rewards of its environment have been built bottom-up, through evolution.

Harm told me: 'A lot of people talking about AI are too optimistic, they underestimate how hard it is to build systems.' Based on his experience of developing *Ms Pac-Man* and other machine-learning systems, he felt we were really far away from a general form of AI.

Even if we can create full bacterial intelligence, Harm was sceptical about how much further we can go. He said: 'Humans are really good at reusing what we learn in doing one task for a different related task; our current state-of-the-art algorithms are terrible in this.'

Both Harm at Microsoft and Tomas Mikalov at Facebook saw a risk in giving neural networks fancy names and making big claims.

The founder of the company Harm now works for seems to agree with him. In September 2017, Bill Gates told the *Wall Street Journal* the subject of AI is not something we need to panic about. He said he disagreed with Elon Musk about the urgency of the potential problems.

So if we are currently mimicking a level of 'intelligence' around that of a tummy bug, why has Elon Musk declared AI such a big concern? Why is Stephen Hawking getting so worried about the predictive power of his speech software? What causes Max Tegmark and his buddies to sit in a row and declare, one after another, their belief that superintelligence is on its way? These are smart people; what is clouding their judgement?

I think there is a combination of factors. One is commercial It doesn't hurt DeepMind to have a bit of buzz around artificial intelligence. Demis Hassabis has toned down the emphasis on 'solving intelligence' his company had when Google first acquired DeepMind, and in recent interviews focuses more on solving mathematical optimisation problems. The work on Go demonstrates that DeepMind has a leading edge on problems like drug discovery and energy optimisation in power networks that require heavy computation to find the best solution out of many available alternatives. Without a bit of hype early on, DeepMind might not have acquired the resources to solve some of these important problems.

Elon Musk hasn't toned down his rhetoric. He appears to be adopting a deliberate position of continually hyping AI to push many of his super-ambitious projects, including self-driving cars. These long-term projects can only succeed if customers are willing to buy into the idea that purchasing the latest Tesla car is a step towards a fantastic future. It is often a far-off dream that drives our desire to find things out about the world.

It isn't money *per se* that motivates the people at DeepMind or Tesla. There is also a genuine feeling of excitement among the researchers. At the turn of the millennium, the idea of general AI seemed to be dead. When neural networks solved the image-classification problem in 2012, it felt like a change was finally coming.

The truth behind current algorithms is often much simpler and more banal than the term 'artificial intelligence' implies. When I looked at the algorithms that try to classify us, I found they were statistical representations of more or less the same things we already know about ourselves. When I had looked at algorithms that tried to influence us, I found out that they were exploiting some very simple aspects of our behaviour to decide what search information to show us and what to try to sell us. Neural networks have cracked a few games, but we can't yet see a way up the next mountain. When Alex and I built our own language bot, it could fool us for a few sentences before revealing itself as a total fake.

At the end of his year-long challenge to build Jarvis, the butler, Mark Zuckerberg put a video up on Facebook demonstrating the results. I have to warn you before you watch his upload that it is pure cringe. We are welcomed into Mark's bedroom as he wakes up in his trademark grey T-shirt. Jarvis informs Mark about his daily meetings and reports that he has been entertaining his one-year-old daughter with a 'Mandarin lesson' since she woke up. When Mark goes down to the kitchen, the toaster is already on, having anticipated that it is time for his daily slice. Then there is a ring at the front door entry system, caused by his face recognition software picking up his parents' faces as they approached. The day ends with him and his wife interacting with a sound system that plays them some mood music.

Zuckerberg was very candid in a blog post about what Jarvis can and can't do.[17] In fact, his challenge of programming a butler mirrors many of the challenges I have faced when dissecting the algorithms used in society. His final product is a masterclass in practical application of the techniques we have seen in this book: face and voice recognition through convolutional neural networks, a form of regression model used to predict when he wants some toast or his daughter wants to learn Mandarin and 'also like' algorithms for choosing music that the whole family want to listen to. Facebook has developed a library of programming routines that helps its employees, and in this case Mark Zuckerberg, turn these algorithms into applications.

When I get over the corniness of his video, and delve more deeply into his blog post, I realise that Zuckerberg is pretty smart (I know I might be a bit late on this one). He makes the grade as a data alchemist. While his fellow Silicon Valley colleagues are trying to prove their intellectual credentials by sitting on panel groups hosted by theoretical physicists,

Mark is getting stuck into programming interfaces and working out how to make the most of the tools his company has created. The conclusions he draws make sense: 'AI is closer to being able to do more powerful things than most people expect – driving cars, curing diseases, discovering planets, understanding media. Those will each have a great impact on the world, but we're still figuring out what real intelligence is.'

There are amazing possibilities to develop products and services using the algorithms we have looked at in this book. These algorithms will continue to change our homes, our workplaces and the way we travel, but they are a long way away from general AI. The techniques are giving our toasters, our home stereos, our offices and our cars a form of bacteria-like intelligence. These algorithms have the potential to reduce the menial tasks we have to do, but they won't be anything like humans.

Kathleen Richardson, professor of ethics and culture of robots and AI, at De Montfort University in Leicester, calls the recent progress made in algorithms 'advertising intelligence' as opposed to 'artificial intelligence'. Talking to the BBC World Service, she said that 'really what has happened over the past 10 years is that big corporations have got better at collecting data on consumers and selling products back to these consumers'. Mark Zuckerberg's butler is a perfect example of this. Mark collects a load of data about his Spotify listening, his friends' faces and his daily routine, feeds it into a computer and it helps improve his daily life.

The real danger, according to Kathleen, is not that computer intelligence explodes, but rather that we use the tools we currently have to improve the lives of the few, instead of the lives of the many. There is a danger that we focus on butlers for the super-rich, instead of wider solutions for everyone else.

Max Tegmark and his friends are entitled to sit around speculating about the future of AI, but their discussion should be seen for what it is. It is a bunch of wealthy men with more or less the same socio-economic background, similar education and work experience who want to argue with each other about science fiction. When I discussed these issues with Olle Häggström I fell into the same trap.

Back in the real world, humans will be the only form of human-like intelligence for a long time to come. The real question is whether we will use the algorithms we already have to help a broader society, or to service the needs of the few? I know which of these two options I would prefer.

CHAPTER EIGHTEEN

Back to Reality

I am standing in a small group at a party.
I am just waiting for it to happen.

That same thing that has happened to me at almost every social gathering for the past year. A period during which I have spent most of my time locked in my office: coding algorithms, fitting statistical models, writing about my results and interviewing engineers and scientists over Skype.

There is small talk. I listen politely. Then someone brings it up. It is not the same thing each time. The stories are slightly different, but the theme is the same.

'The danger with Facebook is that it controls what people see,' he says. 'I read about one study they did where they only showed people negative posts and they got depressed. It can control our emotions and these negative vibes spread like a virus.'

I remain quiet and let someone else speak. 'Yes. They profit from selling their data. It's very likely that Trump won the US election because of this company from Oxford or somewhere that downloaded everyone's profile,' she says.

'You see, the problem is that their algorithms have been trained on fake news,' says the next one. 'It's the same with Google. They built some computer to understand language and it started saying it hated Muslims.'

'I know,' the first man returns to the conversation, 'and this is just the start. The scientists reckon that in 20 years time there will be a computer with human-like intelligence that will decide to turn us into paper clips. I read it in Elon Musk's book.'

It is then that I explode. 'First of all,' I say, 'the Facebook study showed that people who were bombarded with negative news had a propensity to write one more negative word per month, a miniscule but statistically significant effect.

Secondly, the company, called Cambridge Analytica, had nothing to do with Trump's election success. Their CEO has made a series of unverifiable claims about what they can do. Thirdly, yes, computers trained on our language do make sexist and racist associations but that's because our society is implicitly prejudiced. Most Google searches give you, and the rest of the population, the blandest, most accurate results imaginable. The biggest problem with the service is that it is overwhelmed with meaningless links trying to get you to go to Amazon and buy more crap. And finally, there have been some interesting developments in neural networks recently, but the question of whether we can create a general AI remains wide open. We can't even get a computer to learn to play *Ms Pac-Man* properly.'

Everyone looks at me. 'Oh, and Elon Musk is an idiot,' I add.

I hate myself. I hate the boring arsehole that I have become. This boring idiot who has read all the scientific papers, who has to spoil everything with details and caveats. I don't even have that much against Elon Musk. He is just doing his job. I added that, so the conversation would return to a modicum of light-heartedness.

I know that I have misjudged the situation. I am being pedantic and petty. Beyond the small inaccuracies in the details, the people I am talking to have genuine worries about how society is changing. This is why they are talking about Facebook, Cambridge Analytica and artificial intelligence.

Directly after my visit to Google, over a year ago now, I had felt the same way they do. I had felt afraid about the future that mathematicians and computer scientists were creating for us.

I now understand that algorithms are not scary in the way I had thought they might be. It is sad that algorithms haven't solved the problems our society has with sexism and racism, but they haven't made them worse, either. They have highlighted issues about bias that we all need to work harder to solve. It is frightening that SCL group can start a company like Cambridge Analytica that markets itself as targeting personalities, when it doesn't have the tools or the data to do so, but that's global capitalism for you. Either accept it, or make sure you attack the system and not its lies and deceptions. It's depressing that Facebook contains so much fake news and Twitter is full of troll bots, but it's nice to know that hardly anyone is listening to them. It's worrying that success online is only partially related to talent, but it is a consolation when we fail. It's a bit of a bummer that you need to be good-looking to hook up on Tinder, but that's not really a big change for you, is it?

Understanding algorithms allows us to understand future scenarios a little bit better. If you can understand how algorithms work today, then it is easier to judge which scenarios are realistic and which are not. The biggest risk I had found from algorithms was when we failed to think rationally about their effects, when we got carried away in science fiction scenarios.

I am conscious now that my journey, as I stand here after my failed attempt to respond to what was just a bit of 'have a go at Facebook' banter and 'what does it all mean' artificial intelligence speculation, has made me a not very interesting person to talk to.

Luckily, I am married. My wife starts talking about *Pokémon Go*. Lovisa sometimes cycles off to meet 25-year-old men in parks to hunt down Pokémon. And now she tells the group about how they gathered last week in order to try to capture Moltres: my wife, the young men in jogging bottoms, parents in their 30s with babies in prams, a gang of kids who happened to be passing by and an elderly couple who take this route every day as part of their daily Pokémon hunt. She got Moltres. Most of them did, but one of the younger kids started crying when the Pokémon hopped out of the ball and flew away. The elderly couple comforted the young boy. They'll try again tomorrow.

'You should work out how the probability of capturing Moltres changes with the number of poké-balls I throw,' Lovisa suggests. The other people in the group I am standing with now nod in agreement. A bit too much agreement, I feel. Why has David been wasting his time with algorithmic pedantry when he could be enlightening us all with the game theory of *Pokémon Go*?

I don't think I'll bother with *Pokémon*. Some things are best left unanalysed. No one wants to know the probability a kid will start crying.[1]

I think instead about those people in the park showing each other their *Pokémon* decks and the blank spaces that indicate the catches they still have to make. People who wouldn't normally meet each other enjoying themselves together.

I think about my 14 and a half-year-old daughter, Elise. She went off the other week to meet a friend from a chat group on Snapchat. Lovisa and I had been worried at first – might this online 'friend' turn out to be a 40-year-old paedophile? Our concerns were unjustified. Elise met up with a normal 13-year-old with bright blue hair. This summer she wants to go to visit another friend she has made online who lives in Poland. They often chat on Skype while they do their homework together. Elise's parents will have to think long and hard about whether this will be allowed or not.

My son, Henry, and I had just returned from a trip to Newcastle. Through Twitter, I had got in contact with another dad, Ryan, who like me, trains his son's football team. I'd arranged to take 32 12-year-old boys from Sweden, to play a series of matches against his team.

Ryan also invited me to talk about *Soccermatics* at his work, the Department of Work and Pensions (DWP). The setting – a stern ministerial building on the outskirts of town – was very different from my talk at Google in London the year before. But inside, I found myself in a small room packed with data scientists with the same enthusiasm for visualising and understanding data.

After my talk, Ryan showed me what he and his team (the work one, that is) were doing. They want to connect local councils in the UK who face similar challenges, such as a big local employer shutting down, so they can build on shared experiences. Ryan's approach used cutting-edge statistics, but focused on helping his colleagues at the DWP who could, in turn, help the people in local communities. 'We want to allow decision-makers to play with the data,' he told me, 'to discover solutions themselves.'

Ryan's view was that we needed to combine algorithmic and human intelligence to solve our problems. Algorithms alone were not enough.

I thought back to something Joanna Bryson had said to me. I had asked her about the rise of algorithms and whether she thought they would become as smart as us. She told me I was asking the wrong question. 'We have already exceeded human intelligence in so many ways,' she said.

Our culture has produced forms of mathematical 'artificial' intelligence for solving problems over thousands of years: from the early geometry of the Babylonians and the Egyptians, through the development of calculus by Newton and Leibniz, through the hand calculator allowing us to perform faster arithmetic, to the modern computer, the connected society and our present-day algorithmic world. We are getting smarter and smarter with the help of mathematical models, and models are improving because we develop them. Algorithms are part of our cultural heritage. We are part of them and they are part of us.

This heritage is being built all around us, in the most unlikely of places, including, it seems, the UK's Department of Work and Pensions.

There are risks with the way we interact online. A lot of it isn't very pretty, but there are such incredible possibilities to be had from working together with algorithms. And for now at least, it is us who control the algorithms and not them that are controlling us. We are shaping mathematics in our own image.

On Sunday morning, a bunch of lads from Gateshead played football against a bunch of lads from Uppsala. It was a sunny day, with good quality teamwork and fair play, and happy parents cheering the boys on. Afterwards we all watched fireworks together at the local rugby club. All because an algorithm had suggested to Ryan and me that we follow each other on Twitter.

I realise I have been daydreaming and tune back in to the conversation. The subject is still Facebook. So I say, 'Did you know that their advertising recommendations include categories for people interested in "toast" and "platypus"?'

'Not on the same sandwich, I hope,' says the woman standing next to me.

Everyone laughs.

I don't feel quite so outnumbered anymore.

Notes

Chapter 1: Finding Banksy

1 www.fortune.com/2014/08/14/google-goes-darpa
2 www.dailymail.co.uk/femail/article-1034538/Graffiti-artist-Banksy-unmasked-public-schoolboy-middle-class-suburbia.html
3 www.bbc.com/news/science-environment-35645371

Chapter 2: Make Some Noise

1 O'Neil, C. 2016. *Weapons of Math Destruction: How big data increases inequality and threatens democracy.* Crown Books.
2 You can see your Google settings at: www.google.com/settings/u/0/ads/authenticated
3 www.thinkwithgoogle.com/intl/en-aunz/advertising-channels/video/campbells-soup-uses-googles-vogon-to-reach-hungry-australians-on-youtube/
4 You can download the plugin at www.noiszy.com
5 *Weapons of Math Destruction* covers examples of these forms of discrimination in detail.
6 Datta, A., Tschantz, M. C. and Datta, A. 2015. 'Automated experiments on ad privacy settings.' *Proceedings on Privacy Enhancing Technologies* 2015, no. 1: 92–112.
7 www.post-gazette.com/business/career-workplace/2015/07/08/Carnegie-Mellon-researchers-see-disparity-in-targeted-online-job-ads/stories/201507080107
8 Amit and his colleagues discuss the potential explanations here: www.fairlyaccountable.org/adfisher/#disc-cause
9 www.propublica.org/article/machine-bias-risk-assessments-in-criminal-sentencing
10 www.propublica.org/article/facebook-lets-advertisers-exclude-users-by-race
11 See www.ajlunited.org

12 Jonathan Albright's work was part of an article in the
    Guardian on autocomplete and fake news: www.theguardian.
    com/technology/2016/dec/04/google-democracy-truth-internet-
    search-facebook
13 Burrell, J. 2016. 'How the machine 'thinks': Understanding
    opacity in machine learning algorithms.' *Big Data & Society* 3,
    no. 1: 2053951715622512.

Chapter 3: The Principal Components of Friendship

1   I collected the last 15 posts from friends' Facebook pages on
    13 December 2016. Most of the posts could be categorised,
    but those that couldn't I left out of the analysis.
2   We start by plotting the first category against all 12 other
    categories. The second category then needs to be plotted
    against all the 11 remaining categories (*i.e.* excluding the
    first). The third needs to be plotted against 10 categories and
    so on. In total there are 12+11+10+ … +1=13 × 12/2=78
    plots to be made.
3   I describe principal component analysis in the main text from
    a geometric perspective. For those of you who have studied
    algebra, I note that the PCA is found by first calculating the
    covariance matrix for the 13 categories. The first principal
    component is then the eigenvector (which is a straight line)
    corresponding to the largest eigenvalue of the covariance
    matrix. The second principal component is the eigenvector
    for the second largest eigenvalue and so on.

Chapter 4: One Hundred Dimensions of You

1   Kosinski, M., Stillwell, D. and Graepel, T. 2013. 'Private
    traits and attributes are predictable from digital records of
    human behavior.' *Proceedings of the National Academy of Sciences*
    110, no. 15: 5802–5.
2   Kosinski, M., Wang, Y., Lakkaraju, H. and Leskovec, J. 2016.
    'Mining big data to extract patterns and predict real-life
    outcomes.' *Psychological methods* 21, no. 4: 493.

3   Costa, P. T. and McCrae, R. R. 1992. 'Four ways five factors
     are basic.' *Personality and individual differences* 13, no. 6:
     653–65.

4   https://research.fb.com/fast-randomized-svd

5   Facebook researchers describe one method they have
     implemented in the following paper: Szlam, A., Kluger, Y.
     and Tygert, M. 2014. 'An implementation of a randomized
     algorithm for principal component analysis.' *arXiv preprint
     arXiv:1412.3510*. The details of how they use these algorithms
     in practice are not generally available.

6   www.npr.org/sections/alltechconsidered/
     2016/08/28/491504844/you-think-you-know-me-
     facebook-but-you-dont-know-anything

7   Louis, J. J. and Adams, P. 'Social dating.' US Patent
     9,609,072, issued 28 March 2017

8   Kluemper, D. H., Rosen, P. A. and Mossholder, K. W. 2012.
     'Social networking websites, personality ratings, and the
     organizational context: More than meets the eye?' *Journal of
     Applied Social Psychology* 42, no. 5: 1143–72.

9   Stéphane, L. E. and Seguela, J. 'Social networking job
     matching technology.' US Patent Application 13/543,616,
     filed 6 July 2012.

10   See the following patents:
     •   Donohue, A. 'Augmenting text messages with emotion
         information.' US Patent Application 14/950,986, filed 24
         November 2015.
     •   MATAS, Michael J., Reckhow, M. W., and Taigman, Y.
         'Systems and methods for dynamically generating emojis
         based on image analysis of facial features.' US Patent
         Application 14/942,784, filed 16 November 2015.
     •   Yu, Y., and Wang, M. 'Presenting additional content
         items to a social networking system user based on
         receiving an indication of boredom.' US Patent 9,553,939,
         issued 24 January 2017.

11   Hibbeln, M. T., Jenkins, J. L., Schneider, C., Joseph, V. and
     Weinmann, M. 2016. 'Inferring negative emotion from
     mouse cursor movements.'

12   Hehman, E., Stolier, R. M. and Freeman, J. B. 2015.
     'Advanced mouse-tracking analytic techniques for enhancing
     psychological science.' *Group Processes & Intergroup Relations*
     18, no. 3: 384–401.

Chapter 5: Cambridge Hyperbolytica

1   The most recent article had been the subject of a legal
     challenge by the company: www.theguardian.com/
     technology/2017/may/14/robert-mercer-cambridge-analytica-
     leave-eu-referendum-brexit-campaigns
2   This is, of course, an example of one such binary statement.
     They are hard to avoid.
3   https://d25d2506sfb94s.cloudfront.net/cumulus_uploads/
     document/0q7lmn19of/TimesResults_160613_
     EUReferendum_W_Headline.pdf
4   The opinion poll does not have the exact age of the people
     interviewed, so in fitting the model I assumed that each
     person had the median reported age. I used logistic regression
     with a logit link function, with the proportions weighted by
     population sizes. To test robustness, I also looked at quadratic
     functions of age, but these did not significantly improve
     model fit.
5   https://d25d2506sfb94s.cloudfront.net/cumulus_uploads/
     document/0q7lmn19of/TimesResults_160613_
     EUReferendum_W_Headline.pdf
6   Skrondal, A. and Rabe-Hesketh, S. 2003. 'Multilevel logistic
     regression for polytomous data and rankings.' *Psychometrika*
     68, no. 2: 267–87.
7   www.ca-political.com/services/#services_audience_segmentation
8   www.theguardian.com/us-news/2015/dec/11/senator-ted-
     cruz-president-campaign-facebook-user-data
9   www.youtube.com/watch?v=n8Dd5aVXLCc
10   Here I include only users with more than 50 'likes' for
     categories that had in total more than 150 'likes'.
11   www.theintercept.com/2017/03/30/facebook-failed-to-
     protect-30-million-users-from-having-their-data-harvested-
     by-trump-campaign-affiliate

12 A series of exposés by Carole Cadwalladr at the *Guardian* and
by reporters at Channel 4 news uncovered a range of alleged
dubious data-management practices at the company. But as
the fallout settled, there was still no evidence that Nix and
his colleagues had created a personality prediction algorithm.
While whistleblower Christopher Wylie claimed that he and
Alex Kogan had helped CA build a 'psychological warfare'
tool, the details of the effectiveness of this weapon itself was
not revealed. The lack of a smoking gun squared with my
own analysis, and with Alex Kogan's assessment – Facebook
data is not yet sufficiently detailed to enable a suitable analysis
to allow the building of adverts targeted to people's
individual disposition. And with Facebook shares dropping
by 7% on news of the mere possibility of 'psychological
warfare' tools, I imagine that the Internet giant will be very
careful about controlling third party access to Facebook user
data in the future.

Chapter 6: Impossibly Unbiased

1   Several other articles have accused COMPAS of being a
black box, but the article – Brennan, T., Dieterich, W. and
Oliver, W. 2004. 'The COMPAS scales: Normative data
for males and females. Community and incarcerated
samples.' Traverse City, MI: Northpointe Institute for
Public Management – gives a pretty much complete
description of how the model is constructed using
regression and singular value decomposition. Page 147 of
this document provides the key equation used for
predicting recidivism.
2   Brennan, T., Dieterich, W. and Ehret, B. 2009. Evaluating
the predictive validity of the COMPAS risk and needs
assessment system. *Criminal Justice and Behavior, 36*(1),
21–40.
3   www.propublica.org/article/machine-bias-risk-assessments-
in-criminal-sentencing
4   Dieterich, W., Mendoza, C. and Brennan, T. 2016.
COMPAS risk scales: Demonstrating accuracy equity and

predictive parity. Technical report, Northpointe, July 2016. www. northpointeinc. com/northpointe-analysis.

5   www.propublica.org/article/how-we-analyzed-the-compas-recidivism-algorithm

6   Corbett-Davies, S., Pierson, E., Feller, A., Goel, S. and Huq, A. (2017) 'Algorithmic decision making and the cost of fairness.' In Proceedings of the 23rd ACM SIGKDD International Conference on Knowledge Discovery and Data Mining, pp. 797–806. ACM.

7   How the job advert is implemented varies between countries. The 'job post' feature is only implemented in the US and Canada, and adverts are checked for potential discrimination. So in practice, it might not be possible for me to get the advert I suggest implemented by Facebook monitors, and you should treat the text in this section as a thought experiment. I did however find a 'helpful' post by a member of the Facebook help team on how to target for demographics, age and gender using a method much less subtle than the one I now describe: www.facebook.com/business/help/community/question/?id=10209987521898034

8   In the full version of my pedantic explanation, I also note that the proportion of women who didn't see the advert but were interested in the job, i.e. 75/825=9.09 per cent, is the same as the proportion of men who were interested and didn't see the advert, i.e. 50/550=9.09 per cent. The numbers in the table are chosen deliberately to create an outcome where the advertising is equally reliable for both groups.

9   Kleinberg, J., Mullainathan, S. and Raghavan, M. 2016. 'Inherent trade-offs in the fair determination of risk scores.' arXiv preprint arXiv:1609.05807.

10   The median age of white defendants in the Broward County data set is 35, while the median age of black defendants is 30.

11   I identified this relationship using a logistic regression model to predict two year recidivism rates for the Broward County data set collected by ProPublica. The dependent variable was whether the defendant reoffended within two years. I split the data into a training set (90 per cent of observations) and a test set (10 per cent of observations). Using the training set,

I found that the logistic model that best predicted recidivism
was based on age (βage=-0.047; P-value $<$ 2e-16) and number
of priors (βpriors= 0.172; P-value $<$ 2e-16), combined with a
constant (βconst=0.885; P-value $<$ 2e-16). This implies that
older defendants are less likely to be arrested for further
crimes, while those with more priors are more likely to be
arrested again. Race was not a statistically significant predictor
of recidivism (in a multivariate model including race, an
African American factor had P-value = 0.427).

12   The most comprehensive of these is Flores, A. W., Bechtel, K.
and Lowenkamp, C. T. 2016. 'False Positives, False Negatives,
and False Analyses: A Rejoinder to Machine Bias: There's
Software Used across the Country to Predict Future
Criminals. And It's Biased against Blacks.' *Fed. Probation* 80: 38.

13   Arrow, K. J. 1950. 'A difficulty in the concept of social
welfare.' *Journal of political economy* 58, no. 4: 328–46.

14   Young, H. P. 1995. *Equity: in theory and practice.* Princeton
University Press.

15   Dwork, C., Hardt, M., Pitassi, T., Reingold, O. and Zemel,
R. 2012. 'Fairness through awareness.' In *Proceedings of the 3rd
Innovations in Theoretical Computer Science Conference*, pp. 214–
26. ACM.

Chapter 7: The Data Alchemists

1   I can recommend an excellent book if you want to learn
more about expected goals and other algorithms in football.
It's called *Soccermatics*.

2   There are a few subtleties here. To make this comparison, let's
assume that we let the football experts watch a film of a match
up until the exact point at which a shot was struck, but don't
show them whether it was a goal or not. We then ask them
whether the shot was a big chance or not, and we take their
assessment as a prediction. Shots labelled big chances are
predicted to be goals, while those not labelled as big chances are
predicted to miss or be saved. In football commentary, this
interpretation of big chances corresponds to the pundits oft-used
pronouncement that a player, 'should have scored from there'.

3   Julia did not have access to the COMPAS algorithm. So she
    reverse-engineered it. The accuracy levels she found in the
    reverse-engineered algorithm generally replicate the levels
    reported by Northpointe in other studies.

4   In order to compare how the accuracy of my model (based
    purely on age and priors) with the COMPAS model (based
    on a raft of metrics and questionnaire answers), I used the test
    set to create an ROC curve. The AUC of my model was
    0.733, which is comparable to the prediction ability of the
    COMPAS model. In other words, a simple model accounting
    for age and priors is (for the Broward County data set) as
    accurate as the COMPAS model.

Chapter 8: Nate Silver vs the Rest of Us

1   www.yougov.co.uk/news/2017/06/09/the-day-after

2   www.nytimes.com/2016/11/01/us/politics/hillary-clinton-
    campaign.html?mcubz=3

3   www.nytimes.com/2016/11/10/technology/the-data-said-
    clinton-would-win-why-you-shouldnt-have-believed-it

4   www.fivethirtyeight.com/features/the-media-has-a-
    probability-problem

5   www.cafe.com/carl-digglers-super-tuesday-special

6   Tetlock, P. E. and Gardner D. 2016. *Superforecasting: The art
    and science of prediction*. Random House.

7   This figure is taken from a project at The Data Face based on
    data collected by Good Judgement Open: www.thedataface.
    com/good-judgment-open-election-2016.

8   PredictIt has only been around for one presidential election,
    but one of its predecessors that works on the same principles,
    Intrade, was used for both the 2008 and 2012 votes. Alex and
    I took these prediction markets' final probability for every
    state over the three elections, and compared them with
    FiveThirtyEight's predictions.

9   Rothschild, D. 2009. 'Forecasting elections: Comparing
    prediction markets, polls, and their biases.' *Public Opinion
    Quarterly* 73, no. 5: 895–916.

10   The Brier score is a single number that accounts both for accuracy and the courage of predictions. This score is calculated from the square of the distance between the prediction and the actual outcome. So if p is the prediction expressed as a probability, and o is the outcome, where $o=1$ if the event occurred and $o=0$ if it didn't then $(p\text{-}o)^2$ is the Brier score for the event. The lower the Brier score the better. A very brave prediction that an event will occur with 100 per cent certainty, that proves to be correct, has the best possible Brier score of zero. On the other hand, if this brave prediction is incorrect it gets the worst possible Brier score of one. Cowardly predictions, that an event has a 50 per cent chance of occurring, get a Brier score of 0.25, irrespective of the outcome.

11   Rothschild, D. 2009. 'Forecasting elections: Comparing prediction markets, polls, and their biases.' *Public Opinion Quarterly* 73, no. 5: 895–916.

12   Most empirical studies are related to the performance of the board of companies, a task closely related to predicting the future of the business. See for example: Ali, M., Lu Ng, Y. and Kulik, C. T. 2014. 'Board age and gender diversity: A test of competing linear and curvilinear predictions.' *Journal of Business Ethics* 125, no. 3: 497–512. For an overall review of the subject see Page, S. E. 2010. *Diversity and complexity.* Princeton University Press.

13   For more on the difficulties of beating the wise crowd see Buchdahl, J. 2016. *Squares & Sharps, Suckers & Sharks: The Science, Psychology & Philosophy of Gambling.* Vol. 16. Oldcastle Books.

Chapter 9: We 'Also Liked' the Internet

1   This model is usually called 'preferential attachment' in mathematical literature, but has a variety of names reflecting the variety of times it has been discovered. The best mathematical description about how it is used and works can be found in Mark Newman's article on power laws: Newman, M. E. J. 2005. 'Power laws, Pareto

distributions and Zipf's law.' *Contemporary Physics* 46, no. 5: 323–51.

2   The detailed description of the 'also liked' model is as follows. On each step of the model a new customer arrives and looks at the site of their favourite author. The probability that a particular author, *i*, is this new customer's favourite depends on previous sales is given by:

$$\frac{n_i + 1}{N + 25}$$

where n_i is the number of books sold by author *i* and $N=\Sigma_{i=1}^{25}n_i$ is the total number of books sold by all authors. Once at their favourite author's site, the customer decides on a new book to buy based on the number of 'also liked' common purchases between their favourite author and other authors. So the probability they buy a book by author *j* is

$$\frac{c_{ij} + 1}{n_i + 25}$$

where c_{ij} is the number of books sold by author *i* through a connection to author *j*. We then increment c_{ij}, c_{ji} and n_j by one to reflect the new purchase, and the next customer arrives. In Figures 9.1 (a) and (b) the radius of the circle next to the author is proportional to n_i and the thickness of the link between the authors is c_{ij}.

3   Lerman, K. and Hogg, T. 2010. 'Using a model of social dynamics to predict popularity of news.' *Proceedings of the 19th international conference on World Wide Web*, pp. 621–30. ACM.

4   Burghardt, K., Alsina, E. F., Girvan, M., Rand, W. and Lerman, K. 2017. 'The myopia of crowds: Cognitive load and collective evaluation of answers on Stack Exchange.' *PloS one* 12, no. : e0173610.

5   Salganik, M. J., Dodds, P. S. and Watts, D. J. 2006. 'Experimental study of inequality and unpredictability in an artificial cultural market.' *Science* 311, no. 5762: 854–6. Salganik, M. J. and Watts, D. J. 2008. 'Leading the herd astray: An experimental study of self-fulfilling prophecies in an artificial cultural market.' *Social psychology quarterly* 71, no. 4: 338–55.

6   Muchnik, L., Aral, S. and Taylor, S. J. 2013. 'Social influence
     bias: A randomized experiment.' *Science* 341, no. 6146:
     647–51.

7   www.popularmechanics.com/science/health/a9335/
     upvotes-downvotes-and-the-science-of-the-reddit-
     hivemind-15784871

8   The PageRank algorithm works by assuming we browse the
     Internet by following links at random. The page with the
     highest page rank is the one that would be visited most as
     a result of this random browsing. For a mathematical
     description of the Google algorithm see Franceschet,
     Massimo. 2011. 'PageRank: Standing on the shoulders
     of giants.' *Communications of the ACM* 54, no. 6: 92–101.

9   This hypothesis hasn't been tested explicitly, but work
     showing that the average grade on Goodreads declines when
     books have received prestigious awards supports the
     hypothesis. See Kovács, B. and Sharkey, A. J. 2014. 'The
     paradox of publicity: How awards can negatively affect the
     evaluation of quality.' *Administrative Science Quarterly* 59.1:
     1–33.

Chapter 10: The Popularity Contest

1   Giles, J. 2005. 'Science in the web age: Start your engines.'
     Nature, 438 (7068), 554–5.

2   www.backchannel.com/the-gentleman-who-made-scholar-
     d71289d9a82d#.ld8ob7qo9

3   Eom, Y-H. and Fortunato, S. 2011. 'Characterizing and
     modeling citation dynamics.' *PloS one* 6, no. 9: e24926.

4   Lets just look at this mathematically. The plot shows the
     probability, p, that an article is cited more than n times or
     more. Since the data is shown on a log scale and there is a
     straight line relationship between them, we have

$$\log(p) = \log(k) - a \log(n)$$

where a is the slope of the line and k is a constant. From this
we see that

$$p = kn^{-a}$$

which is the power law relationship.

5   May, R. M. 1997. 'The scientific wealth of nations.' *Science*, *275* (5301), 793–6.

6   Petersen, A. M., et al. 2014. 'Reputation and impact in academic careers.' *Proceedings of the National Academy of Sciences* 111.13: 15316–21

7   Higginson, A. D. and Munafò, M. R. 2016. 'Current incentives for scientists lead to underpowered studies with erroneous conclusions.' *PLoS biology*. 14: 11, e2000995.

8   Penner, O., Pan, R. K., Petersen, A. M., Kaski, K. and Fortunato, S. 2013. 'On the predictability of future impact in science.' *Scientific Reports* 3.

9   Acuna, D. E., Allesina, S. and Kording, K. P. 2012. 'Future impact: Predicting scientific success.' *Nature* 489, no. 7415: 201–2.

10   Sinatra, R., Wang, D., Deville, P., Song, C. and Barabási, A-L. 2016. 'Quantifying the evolution of individual scientific impact.' *Science* 354, no. 6312: aaf5239.

11   Tyson, G., Perta, V. C., Haddadi, H. and Seto, M. C. 2016. 'A first look at user activity on Tinder.' *Advances in Social Networks Analysis and Mining (ASONAM), 2016 IEEE/ACM International Conference* pp. 461–6. IEEE.

12   www.collective-behavior.com/the-unstable-dating-game

Chapter 11: Bubbling Up

1   In his book and TED Talk on the filter bubble, Eli Pariser revealed the extent to which our online activities are personalised. Pariser, Eli. 2011. *The Filter Bubble: How the new personalized web is changing what we read and how we think.* Penguin. Google, Facebook and other big Internet companies store data documenting the choices we make when we browse online and then use it to decide what to show us in the future.

2   https://newsroom.fb.com/news/2016/04/news-feed-fyi-from-f8-how-news-feed-works

3   www.techcrunch.com/2016/09/06/ultimate-guide-to-the-news-feed

4   In the model, the probability a user chooses the *Guardian* at
    time t is equal to

$$\frac{G(t)^2+K^2}{G(t)^2+K^2+T(t)^2+K^2}.$$

    where $G(t)$ is the number of times the user has already chosen
    the *Guardian* and $T(t)$ is the number of times the user has
    chosen the *Telegraph*. The square terms here captures the
    combined feedback of (your interest in newspaper) ×
    (closeness to friend sharing article). In the simulation shown
    in the figure, the constant $K=5$.

5   Del Vicario, M., et al. 2016. 'The spreading of
    misinformation online.' *Proceedings of the National Academy of
    Sciences* 113.3: 554–9.

6   www.pewresearch.org/fact-tank/2014/02/03/6-new-
    facts-about-facebook

7   The scientists also performed a control experiment where no
    message was shown. The result in terms of voting numbers
    was very similar to the message without the pictures of
    friends, suggesting that non-social messages are not
    particularly useful. Full details can be found in: Bond,
    R. M., et al. 2012. 'A 61-million-person experiment in social
    influence and political mobilization.' *Nature* 489.7415: 295–8.

Chapter 12: Football Matters

1   A series of papers by Huckfeldt reveals a (what was for me)
    surprising high diversity in political discussions. Huckfeldt,
    R., Beck, P. A., Dalton, R. J., & Levine, J. 1995. 'Political
    environments, cohesive social groups, and the communication
    of public opinion.' *American Journal of Political Science*,
    1025–54.

2   Huckfeldt, R. 2017. 'The 2016 Ithiel de Sola Pool Lecture:
    Interdependence, Communication, and Aggregation:
    Transforming Voters into Electorates.' *PS: Political Science &
    Politics* 50.1: 3–11.

3   DiFranzo, D. and Gloria-Garcia, K. 2017. 'Filter bubbles and
    fake news.' *XRDS: Crossroads, The ACM Magazine for Students*
    23, no. 3: 32–5.

4   Jackson, D., Thorsen, E. and Wring, D. 2016. 'EU
     Referendum Analysis 2016: Media, Voters and the
     Campaign.' www.referendumanalysis.eu

5   The number six shouldn't be taken literally. The most recent
     measurements on Facebook put the average degrees of separation
     at 4.57. On Twitter the degrees of separation is even lower.

6   Johansson, J. 2017. 'A Quantitative Study of Social Media
     Echo Chambers', master's thesis, Uppsala University.

7   These results are taken from Kulshrestha, J., Eslami, M.,
     Messias, J., Zafar, M. B., Ghosh, S., Gummadi, K. P. and
     Karahalios, K. 2017. 'Quantifying search bias: Investigating
     sources of bias for political searches in social media.' *arXiv
     preprint arXiv:1704.01347*.

Chapter 13: Who Reads Fake News?

1   www.buzzfeed.com/craigsilverman/viral-fake-election-
     news-outperformed-real-news-on-facebook?utm_term=.
     rrw0PaV3wP#.qr7rjqJeAj

2   Allcott, H. and Gentzkow, M. 2017. Social media and fake
     news in the 2016 election. No. w23089. National Bureau of
     Economic Research.

3   For more information see www.washingtonpost.com/news/
     the-fix/wp/2016/11/14/googles-top-news-link-for-final-
     election-results-goes-to-a-fake-news-site-with-false-numbers

4   According to SharedCount in January 2017, there had been
     530,858 links to the page on Facebook.

5   Franks, N. R., Gomez, N., Goss, S. and Deneubourg, J-L.
     1991. 'The blind leading the blind in army ant raid patterns:
     testing a model of self-organization (Hymenoptera:
     Formicidae).' *Journal of Insect Behavior* 4, no. 5: 583–607.

6   Ward, A. J. W., Herbert-Read, J. E., Sumpter, D. J.T. and
     Krause, J. 2011. 'Fast and accurate decisions through
     collective vigilance in fish shoals.' *Proceedings of the National
     Academy of Sciences* 108, no. 6: 2312–5.

7   Biro, D., Sumpter, D. J.T., Meade, J. and Guilford, T. 2006. 'From
     compromise to leadership in pigeon homing.' *Current Biology*
     16, no. 21: 2123–8.

8   Cassino, D. and Jenkins, K. 2013. 'Conspiracy Theories
     Prosper: 25 per cent of Americans Are 'Truthers'.' Press
     release – http://publicmind.fdu.edu/2013/outthere.

9   See for example this response by Facebook. http://newsroom.
     fb.com/news/2016/12/news-feed-fyi-addressing-hoaxes-
     and-fake-news

10   Loader, B. D., Vromen, A. and Xenos, M. A. 2014. 'The
     networked young citizen: social media, political participation
     and civic engagement': 143–50.

Chapter 14: Learning to be Sexist

1   There are a wide range of observations and experiments
     confirming these biases. For example, in her 2013 book *The
     American Non-Dilemma: Racial inequality without racism*, Nancy
     DiTomaso documents extensive in-group biases among white
     Americans.

2   Lavergne, M. and Mullainathan, S. 2004. 'Are Emily and
     Greg more employable than Lakisha and Jamal? A field
     experiment on labor market discrimination.' *The American
     Economic Review* 94, no. 4: 991–1013.

3   Moss-Racusin, C. A., Dovidio, J. F., Brescoll, V. L., Graham,
     M. J. and Handelsman, J. 2012. 'Science faculty's subtle
     gender biases favor male students.' *Proceedings of the National
     Academy of Sciences* 109, no. 41: 16474–9.

4   Greenwald, A. G., McGhee, D. E. and Schwartz, J. L. K.
     1998. 'Measuring individual differences in implicit cognition:
     the implicit association test. *Journal of Personality and Social
     Psychology* 74, no. 6: 1464.

5   Take the test yourself at www.implicit.harvard.edu/implicit/
     takeatest

6   Answer: a steer.

7   www.theguardian.com/technology/2016/dec/04/google-
     democracy-truth-internet-search-facebook

8   www.theguardian.com/technology/2016/dec/05/google-
     alters-search-autocomplete-remove-are-jews-evil-suggestion

9   Tosik, M., Hansen, C. L., Goossen, G. and Rotaru, M. 2015.
     'Word embeddings vs word types for sequence labeling: the
     curious case of CV parsing.' In *VS@ HLT-NAACL*, pp. 123–8.

10 Bolukbasi, T., Chang, K-W, Zou, J. Y., Saligrama, V. and
    Kalai, A. T. 2016. 'Man is to computer programmer as
    woman is to homemaker? Debiasing word embeddings.' In
    Advances in Neural Information Processing Systems, pp. 4349–57.

11 For a review see Hofmann, W., Gawronski, B.,
    Gschwendner, T., Le, H. and Schmitt, M. 2005. 'A meta-
    analysis on the correlation between the implicit association
    test and explicit self-report measures.' *Personality and Social
    Psychology Bulletin* 31, no. 10: 1369–85.

Chapter 15: The Only Thought Between the Decimal

1 A similar gate can be found in quantum computing and is
   known as the Hadamard gate.
2 To implement this model I used the neural network demo
   program at https://lecture-demo.ira.uka.de. Here you can
   play about with a range of tools for neural networks.
3 www.tensorflow.org
4 Two good descriptions of recurrent neural networks can be
   found at http://colah.github.io/posts/2015-08-Understanding-
   LSTMs and http://karpathy.github.io/2015/05/21/rnn-
   effectiveness
5 Vinyals, O. and Le, Q. 2015. 'A neural conversational model.'
   arXiv preprint arXiv:1506.05869.
6 Sutskever, I., Vinyals, O. and Le, Q. 2014. 'Sequence to
   sequence learning with neural networks.' *Advances in Neural
   Information Processing Systems*, pp. 3104–12.
7 Mikolov, T., Joulin, A. and Baroni, M. 2015. 'A roadmap
   towards machine intelligence.' *arXiv preprint arXiv:1511.08130.*
8 I have also taken the liberty of correcting punctuation
   mistakes.

Chapter 16: Kick Your Ass at *Space Invaders*

1 http://static.ijcai.org/proceedings-2017/0772.pdf
2 Mnih, V., Kavukcuoglu, K., Silver, D., Rusu, A. A., Veness,
   J., Bellemare, M. G., Graves, A., *et al.* 2015. 'Human-level
   control through deep reinforcement learning.' *Nature* 518, no.
   7540: 529–33.

3   The dataset used is described here: www.image-net.org/
    about-overview

4   www.qz.com/1034972/the-data-that-changed-the-direction-
    of-ai-research-and-possibly-the-world

5   Details of the networks used and the winners can be found
    here: http://cs231n.github.io/convolutional-networks/#case

6   http://selfdrivingcars.mit.edu

7   https://arxiv.org/pdf/1609.08144.pdf

8   https://arxiv.org/pdf/1708.04782.pdf

Chapter 17: The Bacterial Brain

1   www.youtube.com/watch?v=OFBwz4R6Fi0&feature=
    youtu.be

2   www.haggstrom.blogspot.se/2013/10/guest-post-by-david-
    sumpter-why

3   Häggström, O. 2016. *Here Be Dragons: Science, Technology and
    the Future of Humanity*. Oxford University Press.

4   www.youtube.com/watch?v=V0aXMTpZTfc

5   Hassabis, D., Kumaran, D., Summerfield, C. and Botvinick,
    M. 2017. 'Neuroscience-inspired artificial intelligence.'
    Neuron 95, no. 2: 245–58.

6   Kaminski, J. and Nitzschner, M. 2013. 'Do dogs get the
    point? A review of dog–human communication ability.'
    Learning and Motivation 44, no. 4: 294–302.

7   The text that follows is based on the review Chittka, Lars.
    2017. 'Bee cognition.' *Current Biology* 27, no. 19: R1049–53.

8   Loukola, O. J., Clint P. J., Coscos, L., and Chittka, Lars.
    'Bumblebees show cognitive flexibility by improving on an
    observed complex behavior.' Science 355, no. 6327 (2017):
    833–836.

9   Bianconi, E., Piovesan, A., Facchin, F., Beraudi, A., Casadei,
    R., Frabetti, F., Vitale, L., et al. 2013. 'An estimation of the
    number of cells in the human body.' *Annals of Human Biology*
    40, no. 6: 463–71.

10 Herculano-Houzel, S. 2009. 'The human brain in numbers: a
    linearly scaled-up primate brain.' *Frontiers in Human
    Neuroscience* vol. 3.

11   Scholz, M., Dinner, A. R., Levine, E. and Biron, D. 2017.
     'Stochastic feeding dynamics arise from the need for
     information and energy.' *Proceedings of the National Academy of
     Sciences* 114, no. 35: 9261–6.

12   Duisseau, R. P., Vogel, D. and Dussutour, A. 2016.
     'Habituation in non-neural organisms: evidence from slime
     moulds.' In *Proc. R. Soc. B*, vol. 283, no. 1829, p. 20160446.
     The Royal Society.

13   I look at this in more detail in this paper: Ma, Q., Johansson,
     A., Tero, A., Nakagaki, T. and Sumpter, D. J. T. 2013.
     'Current-reinforced random walks for constructing transport
     networks.' *Journal of the Royal Society Interface* 10, no. 80:
     20120864.

14   Baker, M. D. and Stock, J. B. 2007. 'Signal transduction:
     networks and integrated circuits in bacterial cognition.'
     Current Biology 17, no. 23: R1021–4.

15   Turing, A. M. 1950. 'Computing machinery and
     intelligence.' *Mind* 59, no. 236: 433–60.

16   I looked at one such example in the following article:
     Herbert-Read, J. E., Romenskyy, M. and Sumpter, D. J. T.
     2015. 'A Turing test for collective motion.' *Biology letters* 11,
     no. 12: 20150674.

17   www.facebook.com/zuck/posts/10154361492931634

Chapter 18: Back to Reality

1   Although you can find this on Reddit, of course: www.reddit.
     com/r/TheSilphRoad/comments/6ryd6e/cumulative_
     probability_legendary_raid_boss_catch

Acknowledgements

Thank you to all the people who I interviewed or answered my questions over email for this book. These include, but are not limited to, Adam Calhoun, Alex Kogan, Amit Datta, Angela Grammatas, Anupam Datta, Bill Dieterich, Bob Huckfeldt, Brian Connelly, 'CCTV Simon', David Silver, Emilio Ferrara, Garry Gellade, Glenn McDonald, Harm van Seijen, Hunt Allcott, Joanna Bryson, Johan Ydring, Julia Dressel, Katie Genter, Kathleen Richardson, Kristina Lerman, Marc Keuschnigg, Matthew Gentzkow, Michael Wood, Michela Del Vicario, Mona Chalabi, Olle Häggström, Oriol Vinyals, Santo Fortunato, Tassos Noulas, Tim Brennan, Tim Laue and Tomas Mikolov. I learnt so much from all of you. Thank you very much.

I want to say a special thanks to both Renaud Lambiotte and Michal Kosinski for long conversations I had with both of you as I started this book. You provided me with a proper rational framework for reasoning about the dangers, real and imaginary, of algorithms. This book would have been not nearly as good without that input.

Thanks to Mum and Dad for reading the book so thoroughly and for all the comments. Even now I am properly grown up, I still feel you support me so much.

At Bloomsbury, I would like to thank Rebecca Thorne for the trip to Google that started this book; Anna MacDiarmid for all her input and encouragement and Jim Martin for his faith in me. I'm not sure how we can turn this book into an excuse to see the Pars together, but I'll try to think of something.

Thank you to Emily Kearns for careful copyediting and numerous improvements made to the text.

A special thanks to Alex Szorkovszky for his work on the Tolstoy generator, measuring the accuracy of election predictions and dissection of Tinder. Thanks also for putting up with my long uncontrollable rants about the pros and cons of algorithms. This last piece of thanks should also be extended to my research group in Uppsala. Thank you, in particular, to my PhD students (Ernest, Björn and Linnea) who have been so patient with me during the writing of this book.

Thank you to Anna Baddeley for commissioning and giving feedback on *Outnumbered* articles for *Economist 1843*. I learnt a lot from you about the level and balance required when writing about maths.

Thank you to Joakim Johannson for doing his master's project on Twitter and helping me understand echo chambers on the platform.

The idea for *Outnumbered* evolved from a conversation (the day after I visited Google) with my agent Chris Wellbelove about writing a book called *Maths is Evil*. Maths didn't turn out to be quite as evil as we maybe saw it that day but, as ever, I am grateful for the clear way of thinking you provide. I have learnt so much from you.

The title itself is due to my wife Lovisa, who told it to me reflexively when I told her what I wanted to write. Thank you for this and so much more. I enjoy every day of my life with you.

This book would not have been possible without the help of my children, Elise and Henry. Clearly, you both understand everything to do with our life online a lot better than I do, so thanks for patiently explaining it to me. And for being the best kids ever.

Index